PRINCIPLES AND PRACTICE OF ENGINEERING
ELECTRICAL AND COMPUTER ENGINEERING
SAMPLE QUESTIONS & SOLUTIONS

PRINCIPLES AND PRACTICE OF ENGINEERING
ELECTRICAL AND COMPUTER ENGINEERING
SAMPLE QUESTIONS & SOLUTIONS

Published by the
National Council of Examiners for Engineering and Surveying®
280 Seneca Creek Road, Clemson, SC 29631   800-250-3196   www.ncees.org

ISBN 1-932613-11-0

Printed in the United States of America

TABLE OF CONTENTS

INTRODUCTION

The National Council of Examiners for Engineering and Surveying (NCEES) has prepared this handbook to assist candidates who are preparing for the Principles and Practice of Engineering (PE) examination in electrical and computer engineering. The NCEES is an organization established to assist and support the licensing boards that exist in all states and U.S. territories. One of the functions of the NCEES is to develop examinations that are taken by candidates for licensure as professional engineers. The NCEES then provides the licensing boards with uniform examinations that are valid measures of minimum competency related to the practice of engineering.

To develop reliable and valid examinations, the NCEES employs procedures using the guidelines established in the *Standards for Educational and Psychological Testing* published by the American Psychological Association. These procedures are intended to maximize the fairness and quality of the examinations. To ensure that the procedures are followed, the NCEES uses experienced testing specialists possessing the necessary expertise to guide the development of examinations using current testing techniques.

Committees composed of professional engineers from throughout the nation prepare the examinations. These licensed engineers supply the content expertise that is essential in developing examinations. By using the expertise of licensed engineers with different backgrounds such as private consulting, government, industry, and education, the NCEES prepares examinations that are valid measures of minimum competency.

LICENSING REQUIREMENTS

ELIGIBILITY
The primary purpose of licensure is to protect the public by evaluating the qualifications of candidates seeking licensure. While examinations offer one means of measuring the competency levels of candidates, most licensing boards also screen candidates based on education and experience requirements. Because these requirements vary between boards, it would be wise to contact the appropriate board. Board addresses and telephone numbers may be obtained by visiting our Web site at www.ncees.org or calling (800) 250-3196.

APPLICATION PROCEDURES AND DEADLINES
Application procedures for the examination and instructional information are available from individual boards. Requirements and fees vary among the boards, and applicants are responsible for contacting their board office. Sufficient time must be allotted to complete the application process and assemble required data.

DESCRIPTION OF EXAMINATIONS

EXAMINATION SCHEDULE

The NCEES PE examination in electrical and computer engineering is offered to the boards in the spring and fall of each year. Dates of future administrations are as follows:

Year	Spring Dates	Fall Dates
2004	April 16	October 29
2005	April 15	October 28
2006	April 21	October 27

You should contact your board for specific locations of exam sites.

EXAMINATION CONTENT

The 8-hour PE examination in electrical and computer engineering is administered in two 4-hour sessions each containing 40 questions. Each question has four answer options. The morning session covers the breadth of electrical and computer engineering, and the afternoon session presents three alternate modules from which the examinee selects one. The examination specifications presented in this book give details of the subjects covered on the examination. The three afternoon modules are:

1. Computers
2. Electronics, Controls, and Communications
3. Power

This book presents a complete sample examination. By illustrating the general content of the topic areas, the questions should be helpful in preparing for the examination. Solutions are presented for all the questions. The solution presented may not be the only way to solve the question. The intent is to demonstrate the typical effort required to solve each question.

No representation is made or intended as to future examination questions, content, or subject matter.

EXAMINATION DEVELOPMENT

EXAMINATION VALIDITY

Testing standards require that the questions on a licensing examination be representative of the important tasks needed for competent practice in the profession as a licensed professional. The NCEES establishes the relationship between the examination questions and tasks by conducting an analysis using licensed practitioners who identify the duties performed by an engineer. This information is used to develop an examination content outline that guides the development of job-related questions.

EXAMINATION SPECIFICATIONS

The examination content outline presented in this book specifies the subject areas that were identified for electrical and computer engineering and the percentage of questions devoted to each of them. The percentage of questions assigned to each of the topic areas reflects both the frequency and importance experienced in the practice of electrical and computer engineering.

EXAMINATION PREPARATION AND REVIEW

Examination development and review workshops are conducted frequently by standing committees of the NCEES. Additionally, question-writing workshops are held as required to supplement the bank of questions available. The content and format of the questions are reviewed by the committee members for compliance with the specifications and to ensure the quality and fairness of the examination. These licensed engineers are selected with the objective that they be representative of the profession in terms of geography, ethnic background, gender, and area of practice.

MINIMUM COMPETENCY

One of the most critical considerations in developing and administering examinations is the establishment of passing scores that reflect a standard of minimum competency. The concept of minimum competency is uppermost in the minds of the committee members as they assemble questions for the examination. Minimum competency, as measured by the examination component of the licensing process, is defined as the lowest level of knowledge at which a person can practice professional engineering in such a manner that will safeguard life, health, and property and promote the public welfare.

To accomplish the setting of fair passing scores that reflect the standard of minimum competency, the NCEES conducts passing score studies. At these studies, a representative panel of engineers familiar with the examinee population uses established procedures to set the passing score for the examination. Such procedures are widely recognized and accepted for occupational licensing purposes. The panel discusses the concept of minimum competency and develops a written standard of minimum competency that clearly articulates what skills and knowledge are required of licensed engineers. Following this, the panelists take the examination and then evaluate the difficulty level of each question in the context of the standard of minimum competency.

The NCEES does not use a fixed-percentage pass rate such as 70% or 75% because licensure is designed to ensure that practitioners possess enough knowledge to perform professional activities in a manner that protects the public welfare. The key issue is whether an individual candidate is competent to practice and not whether the candidate is better or worse than other candidates.

The passing score can vary from one administration of the examination to another to reflect differences in difficulty levels of the examinations. However, the passing score is always based on the standard of minimum competency. To avoid confusion that might arise from fluctuations in the passing score, scores are converted to a standard scale that adopts 70 as the passing score. This technique of converting to a standard scale is commonly employed by testing specialists. Some licensing jurisdictions may choose to report examination results on a pass/fail basis and not provide a numeric score.

SCORING PROCEDURES

The examination consists of 80 equally weighted multiple-choice questions. There is no penalty for marking incorrect responses; therefore candidates should answer each question on the examination. Only one response should be marked for each question. No credit is given where two or more responses are marked. The examination is compensatory—poor scores in some topics can be offset by superior performance elsewhere.

The legal authority for making licensure decisions rests with the individual licensing boards and not with the NCEES. Consequently, each board has the authority to determine the passing score for the examination. The NCEES provides each board with a recommended passing score based on the procedures described previously. To assist failing candidates, the NCEES provides the licensing boards with a diagnostic report for each failing candidate that shows the performance on each topic area on the examination.

EXAMINATION PROCEDURES AND INSTRUCTIONS

EXAMINATION MATERIALS

Before the morning and afternoon sessions, proctors will distribute examination booklets containing an answer sheet. You should not open the examination booklet until you are instructed to do so by the proctor. Read the instructions and information given on the front and back covers and enter your name in the upper right corner of the front cover. Listen carefully to all the instructions the proctor reads. The proctor has final authority on the administration of the examination.

The answer sheets for the multiple-choice questions are machine scored. For proper scoring, the answer spaces should be blackened completely. Use only #2 pencils or mechanical pencils with HB lead. Marks in ink or felt-tip pens may not be scanned properly. If you decide to change an answer, you must erase the first answer completely. Incomplete erasures and stray marks may be read as intended answers. One side of the answer sheet is used to collect identification and biographical data that may be used to analyze the performance of the examination. The biographical data has no impact on the examination score. Proctors will guide you through the process of completing this portion of the answer sheet prior to taking the test. This process will take approximately 15 minutes.

STARTING AND COMPLETING THE EXAMINATION

You are not to open the examination booklet until instructed to do so by your proctor. If you complete the examination with more than 30 minutes remaining, you are free to leave after returning all examination materials to the proctor. Within 30 minutes of the end of the examination, you are required to remain until the end to avoid disruption to those still working and to permit orderly collection of all examination materials. Regardless of when you complete the examination, you are responsible for returning the numbered examination booklet assigned to you. Cooperate with the proctors collecting the examination materials. Nobody will be allowed to leave until the proctor has verified that all materials have been collected.

REFERENCES

The PE examination is open-book. Your board determines the reference materials and calculators that will be allowed. In general, you may use textbooks, handbooks, bound reference materials, and a non-communicating, battery-operated, silent, non-printing calculator. Calculating and computing devices having a QWERTY keypad arrangement similar to a typewriter or keyboard are not permitted, nor are communication devices such as pagers and cellular phones. States differ in their rules regarding calculators and references, and you should contact your board for specific advice.

SPECIAL ACCOMMODATIONS

If you require special accommodations in the test-taking procedure, you should communicate your need to your board office well in advance of the day of the examination so that appropriate arrangements may be determined.

UPDATES TO EXAMINATION INFORMATION

For updated exam specifications, errata for this book, and other information about exams, visit the NCEES Web site at www.ncees.org.

EXAMINATION SPECIFICATIONS

ELECTRICAL AND COMPUTER ENGINEERING

The electrical engineering examination is a breadth and depth examination. This means that all examinees work the breadth (AM) exam and one of the three depth (PM) exams. The breadth exam contains questions from the general field of electrical engineering. The depth exams focus more closely on a single area of practice in electrical engineering. The three depth examinations are (1) Computers; (2) Electronics, Controls, and Communications; and (3) Power.

BREADTH (AM) EXAMINATION
EFFECTIVE APRIL 2002

		Approximate Percentage of Examination
I. Basic Electrical Engineering		**45%**
A. Professionalism and Engineering Economics	6%	
1. Engineering Economics		
2. Ethics		
3. Professional Practice		
B. Safety and Reliability	6%	
1. Reliability		
2. Electric Shock and Burns		
3. General Public Safety		
C. Electric Circuits	24%	
1. Ohm's Law		
2. Coulomb's Law		
3. Faraday's Law		
4. Kirchhoff's Laws		
5. Thevenin's Theorem		
6. Norton's Theorem		
7. Superposition		
8. Source Transformation		
9. Sinusoidal Steady State Analysis		
10. Power and Energy Calculations		
11. Transient Analysis		
12. Fourier Analysis		
13. Transfer Functions		
14. Complex Impedance		
15. Laplace Transforms		
16. Mutual Inductance		
D. Electric and Magnetic Field Theory and Applications	3%	
1. Electrostatic Effects		
2. Magnetostatic Fields		
E. Digital Logic	6%	
1. Digital Logic		

II. Electronics, Electronic Circuits and Components		**20%**
A. Components	14%	
1. Solid State Device Characteristics and Ratings		
2. Operational Amplifiers		
3. Transistors		
4. Signal Grounding		
5. Transducers/Sensors		
B. Electrical and Electronic Materials	6%	
1. Conductivity/Resistivity		
2. Thermal Characteristics		
3. Semiconductors		
III. Controls and Communications Systems		**15%**
A. Controls and Communications Systems		
1. System Stability		
2. Frequency Response		
3. Analog Modulation		
4. Frequency Selective Filters		
IV. Power		**20%**
A. Transmission and Distribution	12%	
1. Voltage Regulation		
2. Power Factor Correction		
3. Grounding		
B. Rotating Machines and Electromagnetic Devices	8%	
1. AC and DC Machines		
2. Transformers		
	TOTAL	**100%**

NOTES:

1. The knowledge areas specified under A, B, C, ... etc., are examples of kinds of knowledge, but they are not exclusive or exhaustive categories.
2. The breadth (AM) exam contains 40 multiple-choice questions. Examinee works all questions.

COMPUTERS DEPTH (PM) EXAMINATION
EFFECTIVE APRIL 2002

		Approximate Percentage of Examination
I. General Computer Systems		**10%**
A. Interpretation of Codes and Standards	4%	
1. IEEE Standards		
2. ISO Standards		
B. Microprocessor Systems	6%	
1. Number Systems and Codes		
2. Microprocessor Systems		
a. Components		
b. Control Applications		
c. Math Applications		
d. Programmable Logic Controllers		
e. Real-time Operations		
II. Hardware		**45%**
A. Digital Electronics	16%	
1. Memory Devices		
2. Medium Scale Integration Devices		
3. Programmable Logic Devices and Gate Arrays		
4. Tristate Logic		
5. Digital Electronic Devices		
6. Logic Components		
a. Properties		
b. Fan-In, Fan-Out		
c. Propagation Delay		
7. Large Scale Integration		
8. Analog to Digital and Digital to Analog Conversion		
B. Design and Analysis	19%	
1. Clock Generation/Distribution		
2. Memory Interface		
3. Processor Interfacing		
4. Asynchronous Communication		
5. Metastability		
6. Races and Hazards		
7. State Transition Tables		
8. State Transition Diagrams		
9. Algorithmic State Machine Charts		
10. Timing Diagrams		
11. Synchronous State Machines		
12. Asynchronous State Machines		
13. Pipelining and Parallel Processing		
14. Fault Tolerance		
15. Sampling Theory		
C. Systems	10%	
1. Digital Signal Processor Architecture		
2. Design for Testability		
3. Computer Architecture		
4. Mass Storage Devices		
5. Input/Output Devices		
6. Central Processing Unit Architecture		

III. Software   **35%**
   A. System Software   12%
      1. Computer Security
      2. Real-Time Operating Systems
      3. Error Detection and Control
      4. Drivers
      5. Time Critical Scheduling
   B. Development/Applications   23%
      1. Computer Control and Monitoring
      2. Software Lifecycle
          a. Requirements Definition
          b. Specification
          c. Design
          d. Implementation and Debugging
          e. Testing
          f. Maintenance and Upgrade
      3. Fault Tolerance
      4. Modeling and Simulation
      5. Software Pipelining
      6. Human Interface Requirements
      7. Software Design Methods and Documentation
          a. Structured Programming
          b. Top Down or Bottom Up Programming
          c. Successive Refinement
          d. Programming Specifications
          e. Program Testing
          f. Structure Diagrams
          g. Recursion
      8. Object Oriented Design
      9. Data Structures
          a. Internal
          b. External

IV. Networks   **10%**
   A. Networks
      1. Protocols
          a. TCP/IP
          b. Ethernet
      2. Computer Networks
          a. OSI Model
          b. Network Topology
          c. Network Technology
          d. Network Security

                                              TOTAL   **100%**

NOTES:

1. The knowledge areas specified under A, B, C, ... etc., are examples of kinds of knowledge, but they are not exclusive or exhaustive categories.

2. Each depth (PM) exam contains 40 multiple-choice questions. Examinee chooses **one** depth exam and works all questions in the depth exam chosen.

ELECTRONICS, CONTROLS, AND COMMUNICATIONS
DEPTH (PM) EXAMINATION
EFFECTIVE APRIL 2002

Approximate
Percentage of
<u>Examination</u>

I.   General Electrical Engineering Knowledge   **10%**

    A. Measurement and Instrumentation   4%

        1. Transducer Characteristics

        2. Frequency Response

        3. Quantization

        4. Data Evaluation

        5. Sampling Theory

    B. Interpretation of Codes and Standards   2%

        1. ANSI Standards

        2. NEC (code)

        3. IEEE Standards

        4. FCC Standards

        5. EIA Standards

        6. ISA Standards

        7. ISO Standards

    C. Computer Systems   4%

        1. Programmable Logic Devices

        2. Computer Networks

        3. Number Systems and Codes

        4. Digital Electronic Devices

II.   Electronics   **35%**

    A. Electric Circuit Theory   10%

        1. Small Signal and Large Signal

        2. Active Networks and Filters

        3. Delay

        4. Distributed Parameter Circuits

        5. Nonlinear Circuits

        6. Two Port Theory

        7. Phase Delay

    B. Electric and Magnetic Field Theory and Applications   7%

        1. Microwave Systems

        2. Transmission Line Models

        3. Electromagnetic Fields and Interference

        4. Antennas

        5. Free Space Propagation

        6. Guided Wave Propagation

  C. Electronic Components and Circuits   18%
    1. Programmable Logic Devices
    2. Programmable Gate Arrays
    3. Solid State Power Devices and Applications
    4. Battery Characteristics and Ratings
    5. Power Supplies
    6. Phase Locked Loops
    7. Oscillators
    8. Amplifiers
    9. Modulators and Demodulators
    10. Discrete Components
    11. Diodes
    12. Circuit Protection
    13. Relays and Switches
    14. Logic Components
      a. Properties
      b. Fan In, Fan Out
      c. Propagation Delay
    15. Transistors and Applications

III. Controls   **25%**

  A. Control System Fundamentals   10%
    1. Difference Equations
    2. z - Transform
    3. Frequency Response
    4. Characteristic Equations
    5. Block Diagrams
    6. State Variable Analysis
  B. Control System Design/Implementation   6%
    1. Compensators
    2. Feed Forward
    3. Feedback
    4. Optimal Control Systems
    5. Adaptive Control
    6. Computer Control and Monitoring
    7. Error Actuated Control
    8. Proportional-Integral-Derivative Control
  C. Stability   9%
    1. Stability Analysis and Design
      a. Nyquist Stability
      b. Root Locus
      c. Bode Diagrams
    2. Poles and Zeros
    3. Phase and Gain Margin
    4. Transport Delay

IV. Communications   **30%**

    A. Communications and Signal Processing   15%
        1. Modulation Theory
            a. Linear Modulation
            b. Angle Modulation
            c. Pulse Modulation
        2. Correlation and Convolution
        3. Fourier Transforms
        4. Spectral Properties
        5. Signal Processing
        6. Digital Transmission
        7. Quadrature Amplitude Modulation
        8. Personal Communication System
        9. Spread Spectrum Modulation
        10. Adaptive Filtering
        11. Nyquist Sampling Theorem
    B. Noise and Interference   8%
        1. Signal to Noise Ratio
        2. Quantization Noise
        3. Noise Figure and Temperature
        4. Aliasing
        5. Random Variables
        6. Error Detection and Correction
    C. Telecommunications   7%
        1. Wireless Communications
        2. Compression
        3. Cellular Communications
        4. Optical Communications
        5. Circuit and Packet Switching
        6. Network Distribution Systems
        7. Wireline Communications

                                           TOTAL   **100%**

NOTES:

1. The knowledge areas specified under A, B, C, ... etc., are examples of kinds of knowledge, but they are not exclusive or exhaustive categories.

2. Each depth (PM) exam contains 40 multiple-choice questions. Examinee chooses **one** depth exam and works all questions in the depth exam chosen.

POWER DEPTH (PM) EXAMINATION
EFFECTIVE APRIL 2002

Approximate
Percentage of
Examination

I. General Power Engineering   **15%**

  A. Measurement, Instrumentation and Statistics   5%
     1. Power Metering
     2. Instrument Transformers
     3. Transducers
     4. Frequency Response of Measurement Devices
     5. Data Evaluation
     6. Reliability

  B. Special Applications   2%
     1. Illumination Design
     2. Lightning and Surge Protection

  C. Codes and Standards   8%
     1. ANSI Standards
     2. NEC (code)
     3. IEEE Standards
     4. NEMA Standards
     5. NESC (code)

II. Circuit Analysis   **28%**

  A. Analysis   15%
     1. Short Circuit Analysis
     2. Wye-Delta Transformation
     3. Three-Phase Circuit Analysis
     4. Symmetrical Components
     5. Balanced and Unbalanced Systems
     6. Per Unit Analysis

  B. Devices and Power Electronic Circuits   8%
     1. Solid State Power Device Characteristics and Ratings
     2. Battery Characteristics and Ratings
     3. Power Supplies
     4. Relays and Switches
     5. Power Electronics

  C. Electric and Magnetic Fields and Applications   5%
     1. Transmission Line Models
     2. Mechanical Forces Between Conductors
     3. Electromagnetic Fields, Coupling, and Interference
     4. Electrostatics
     5. Ferroresonance

			Approximate Percentage of Examination

III. Rotating Machines and Electromagnetic Devices — 27%
   A. Rotating Machines — 18%
       1. Synchronous Machines
       2. Induction Machines
       3. DC Machines
       4. Machine Constants and Nameplate Data
       5. Equivalent Circuits
       6. Response Times
       7. Speed-Torque Characteristics
       8. Speed Control
       9. Motor Starting
       10. Variable Speed Drives
       11. Testing
   B. Electromagnetic Devices — 9%
       1. Transformers
       2. Reactors
       3. Magnetic Circuit Theory
       4. Testing

IV. Transmission and Distribution — 30%
   A. System Analysis — 15%
       1. Voltage Drop and Voltage Regulation
       2. Power Factor Correction
       3. Parallel Three-Phase Systems
       4. Surge Protection
       5. Power Quality
       6. Fault Current Analysis
       7. Grounding
       8. Resistance Grounding
       9. Transformer Connections
       10. Models
   B. Power System Performance — 6%
       1. Load Flow
       2. Models
       3. Power System Stability
       4. Voltage Profile
       5. Computer Control and Monitoring
   C. Protection — 9%
       1. Overcurrent Protection
       2. Protective Relaying
       3. Protective Devices
       4. Coordination

TOTAL — 100%

NOTES:

1. The knowledge areas specified under A, B, C, ... etc., are examples of kinds of knowledge, but they are not exclusive or exhaustive categories.

2. Each depth (PM) exam contains 40 multiple-choice questions. Examinee chooses **one** depth exam and works all questions in the depth exam chosen.

ELECTRICAL AND COMPUTER ENGINEERING
BREADTH MORNING SAMPLE QUESTIONS

101. A stack gas analyzer for a power plant can be purchased from any of four possible vendors and installed for a one-time cost as shown in the following table. Monthly benefits and costs for maintenance and engineering support are also shown in the table.

Vendor	One-time Cost ($)	Monthly Benefits and Costs ($)		
		Benefits	Maintenance	Engineering Support
A	24,000	4,000	800	800
B	27,000	4,200	700	1,000
C	30,000	4,500	500	1,200
D	32,000	4,800	400	1,500

Based on the given data and ignoring the effects of depreciation and corporate taxes, which vendor's product would be selected based on a minimum payout-time criterion?

(A) Vendor A
(B) Vendor B
(C) Vendor C
(D) Vendor D

102. Consider the following plan for a computing system to control multiple injection stations.

A small computing system costing $50,000 is installed initially. A second computer is installed 2 years later to handle additional production needs, also at a cost of $50,000. Each unit requires a maintenance cost of $10,000/year of operation.

The computers become obsolete very fast, hence zero salvage is expected by the owners.

Assume a 10% rate of return and a 10% interest rate.

For the first 4 years, the present worth of the investment and maintenance cost is most nearly:

(A) $117,000
(B) $127,000
(C) $137,000
(D) $147,000

103. Two components of a system are connected in parallel. The probability of survival (reliability) R of each of the components making up the system is given below. The combined system works properly if either one of the parallel components functions.

Component Reliability

R_1	0.92
R_2	0.85

The reliability of the combined system (parallel combination) is most nearly:

(A) 0.78
(B) 0.85
(C) 0.92
(D) 0.99

104. To reduce electrical shock hazard, the grounded neutral conductor in a separately derived 3-phase 480Y/277 V distribution system should be electrically bonded to:

(A) the grounding electrode conductor and the equipment grounding conductor at the service equipment or source

(B) a grounding electrode conductor whenever possible at panels on the load side of the service equipment or source

(C) the equipment grounding conductor at each panel board on the load side of the service equipment or source

(D) all of the above

105. A single-phase, 120-VAC branch circuit is properly protected by a circuit breaker and a ground-fault circuit interrupter (GFCI). Which of the following electrical hazards is **LEAST** protected?

    (A)   Short circuit
    (B)   Overload
    (C)   High-resistance fault, phase-to-ground
    (D)   Line-to-neutral shock

106. Tests at two terminals of a linear network produce the following:

    (1)   with the terminals shorted, the current in the short circuit is 3.0 A
    (2)   with a conductance of 0.2 S connected to the terminals, the voltage between the terminals is 5 V.

The linear network may be replaced by a current source in parallel with a conductance of most nearly:

    (A)   1.0 A, 0.2 S
    (B)   3.0 A, 0.4 S
    (C)   2.0 A, 0.6 S
    (D)   3.0 A, 0.6 S

107. A current $i(t) = [5 \cos (\omega t) + 2 \cos (2\omega t)]$ A flows through a 3-Ω resistor. The average power (W) absorbed by the 3-Ω resistor is most nearly:

(A)   44
(B)   74
(C)   87
(D)   147

108. A sinusoidal voltage of $v(t) = [50 \cos(\omega t + 45°) + 28.28 \cos(\omega t)]$ V is applied across a 5-Ω resistor. The average power (W) delivered to the resistor is most nearly:

(A)   330
(B)   530
(C)   660
(D)   1,225

109. The switch in the figure below has been closed for a long time. At $t = 0$ it is opened. The equation for the voltage $v(t)$, for $t > 0$, is of the form $v(t) = A + Be^{-t/\tau}$. The values of A, B, and τ are most nearly:

(A) $A = 4$ V, $B = 2$ V, $\tau = 6$ ms
(B) $A = 0$ V, $B = 4$ V, $\tau = 3$ ms
(C) $A = 0$ V, $B = 6$ V, $\tau = 6$ ms
(D) $A = 0$ V, $B = 4$ V, $\tau = 6$ ms

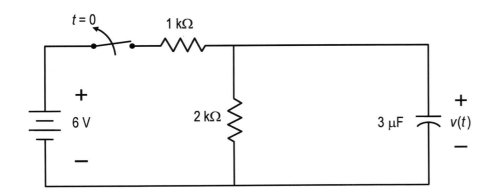

110. A saw-tooth voltage waveform $v(t)$ is shown in the figure below. The Fourier series for this waveform will:

(A)   have a dc component greater than zero
(B)   contain cosines but not sines
(C)   be even
(D)   be odd

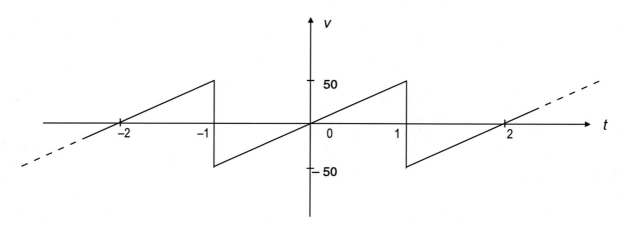

111. The transfer function $v(s)/i(s)$ for the circuit shown in the figure below is:

(A) $$\dfrac{s^2 \dfrac{RL}{C}}{s^2 + (R + R_1)\dfrac{s}{C} + \dfrac{1}{LC}}$$

(B) $$\dfrac{\left(s^2 + \dfrac{R_1}{L}s + \dfrac{1}{LC}\right)sL}{s^2 + \dfrac{(R + R_1)}{L}s + \dfrac{1}{LC}}$$

(C) $$\dfrac{s^2 R}{s^2 + \dfrac{R_1}{L}s + \dfrac{1}{LC}}$$

(D) $$\dfrac{s^2 R}{s^2 + \dfrac{(R + R_1)s}{L} + \dfrac{1}{LC}}$$

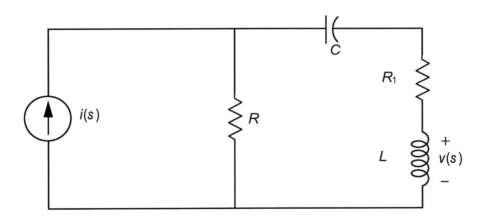

112. Tests at two terminals of a linear network produce the following:

(1)   with the terminals shorted, the current in the short circuit is 5 $\angle 45°$ A$_{rms}$,

(2)   with the terminals open, the voltage between the open terminals is $100\angle 0°$ V$_{rms}$.

This linear network may be replaced by an rms voltage source in series with a complex impedance of most nearly:

(A) $100\angle 45°$ V, $(14 - j14)$ Ω
(B) $100\angle 0°$ V, $(14 + j14)$ Ω
(C) $100\angle 0°$ V, $(14 - j14)$ Ω
(D) $100\angle 0°$ V, $(10 - j10)$ Ω

GO ON TO THE NEXT PAGE

113. **Figure 1** represents equivalent transformations of the components in **Figure 2.** The values of the transformed source and resistor in **Figure 1** are most nearly:

(A) $I_A = -2.5$ A, $G_B = 0.25$ S
(B) $I_A = 2.5$ A, $G_B = 0.25$ S
(C) $I_A = 2.5$ A, $G_B = 1.0$ S
(D) $I_A = 1.25$ A, $G_B = 0.5$ S

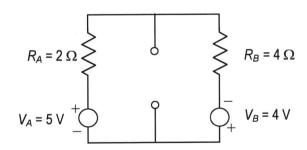

FIGURE 1

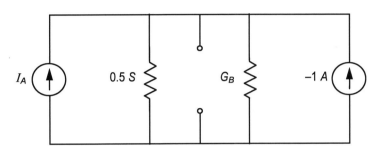

FIGURE 2

114. For the circuit below, the current $I(s)$ is given by the following expression:

(A) $\dfrac{10,000\,s}{(s+1,000)^2}$

(B) $\dfrac{10,000}{(s+1,000)^2}$

(C) $\dfrac{10,000}{(s+1,000)}$

(D) $\dfrac{10,000}{(s-1,000)^2}$

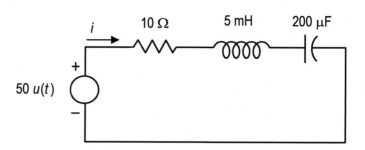

115. A transducer which is modeled as a Norton equivalent is coupled to a load resistance, R_L, by an ideal transformer. To obtain maximum power transfer from the transducer to the load, the turns ratio $N_1{:}N_2$ is most nearly:

(A) 1:6
(B) 2.5:1
(C) 6:1
(D) 36:1

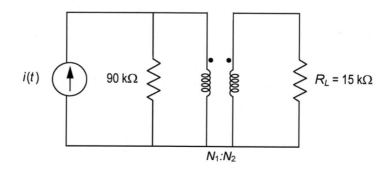

GO ON TO THE NEXT PAGE

116. A dc current of 50 A is flowing in a long isolated wire. The magnitude of the magnetic field intensity, $\overline{H}$ (A/m), 10 cm from the center of the wire is most nearly:

(A)   0.001
(B)   8
(C)   80
(D)   250

117. The output D in the circuit is:

(A) $B(A + C)$
(B) $(A + B)(B + C)$
(C) $\overline{AB} + \overline{BC}$
(D) $(\overline{A} + \overline{B})CB$

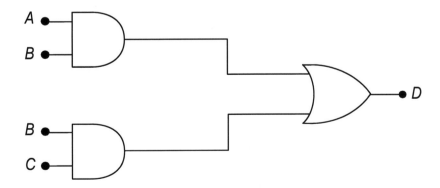

118. The function $F = (\overline{A} + B)(\overline{B} + \overline{C})(A + C)$ can be expressed as the following sum of products:

(A) $F = ABC + A\overline{B}C$

(B) $F = ABC + \overline{A}\,\overline{B}C$

(C) $F = AB\overline{C} + A\overline{B}C$

(D) $F = AB\overline{C} + \overline{A}\,BC$

119. In the circuit below, Switch S1 closes at $t = 0$. The diode D is rated at a maximum peak current of 5 A. Assume that the initial charge on the capacitor is zero. The minimum value of R (Ω) to prevent exceeding the diode maximum peak current is most nearly:

(A)   33
(B)   24
(C)   21
(D)   11

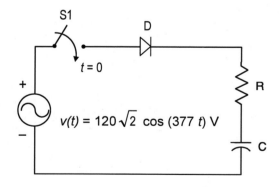

120. The Bode plot for the circuit shown for $A = V_2/V_1$ is most nearly:

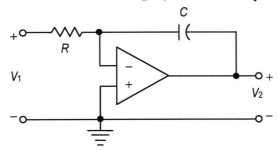

(A)

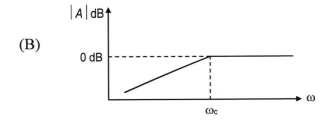

(B)

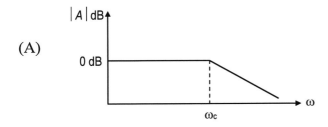

(C)

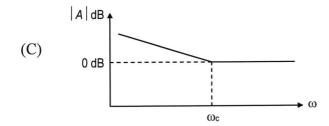

(D)

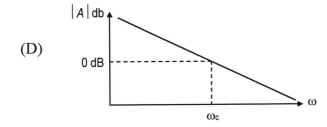

121. A transistor shown in the circuit below has an $h_{FE} = 100$. If $V_C = 10$ V, V_E (V) is most nearly:

(A) 0.7
(B) 2.0
(C) 8.25
(D) 10

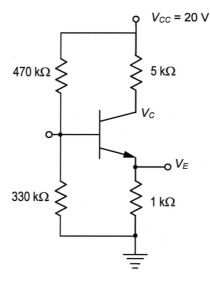

122. A silicon diode is used as a half wave rectifier as shown below. The minimum PIV rating (V) necessary to avoid reverse breakdown is most nearly:

(A) 220
(B) 310
(C) 440
(D) 620

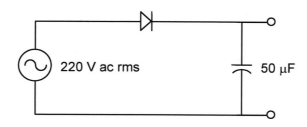

123. A loudspeaker with an electrical impedance of 8 Ω resistive has an efficiency of 15%. If the terminal voltage is 30 V_{rms}, the acoustic output (W) is most nearly:

(A)   4.5
(B)   17
(C)   36
(D)   135

124. The silicon transistor shown has coupling capacitors that may be treated as a short circuit for signals of interest. The source v_i is sinusoidal. The peak value of a sinusoidal signal (V) at v_o that will not be saturated or clipped is most nearly:

(A)   4.6
(B)   5
(C)   5.4
(D)   10

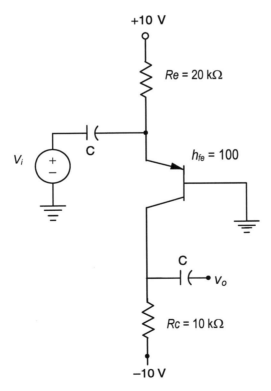

125. When silicon is used as a semiconductor material at room temperature and is a heavily doped n-type, the conductivity of the material is most strongly dependent on:

   (A)   electron concentration
   (B)   hole concentration
   (C)   intrinsic concentration
   (D)   cross-sectional area

126. The zener diode, D_1, has a reverse breakdown voltage of 5.1 V and is ideal. The average value of the output voltage, v_0 (V), is most nearly:

   (A)   0
   (B)   2.5
   (C)   3.5
   (D)   7.0

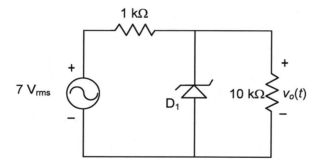

127. For the parameters shown in the figure, the closed-loop damping ratio is most nearly:

(A) 0.40
(B) 0.63
(C) 0.70
(D) 2.82

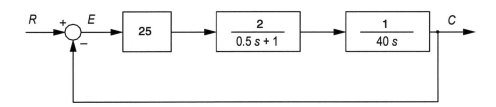

128. At high frequencies, the slope of the Bode plot (dB/decade) for the transfer function $\dfrac{100\, s\, (s+1)}{(s+2)(s+3)(s+4)^2}$ is most nearly:

(A) +20
(B) −20
(C) −40
(D) −80

129. To make a stable discrete-time system more oscillatory, the closed-loop poles in the z-plane must be moved:

(A) closer to the origin
(B) further into the left-half plane
(C) closer to the jω axis
(D) closer to the unit circle

130. The system shown is stable for:

(A) $K > -2$
(B) $-3 < K < 0$
(C) $K < -2$
(D) $-3 < K < -2$

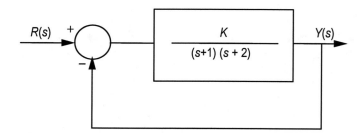

GO ON TO THE NEXT PAGE

131. If a carrier wave of 10 MHz is amplitude modulated by an audio signal of 1 kHz, the resultant signal to be broadcast contains only the:

   (A)   10,001,000-Hz component
   (B)   10,001,000-Hz and 9,999,000-Hz components
   (C)   10,001,000-Hz, 10,000,000-Hz, and 9,999,000-Hz components
   (D)   10,001,000-Hz, 10,000,000-Hz, 9,999,000-Hz, and 1,000-Hz components

132. For the system response in the figures the phase margin of the system is most nearly:

(A) –270°
(B) –90°
(C)   0°
(D)   90°

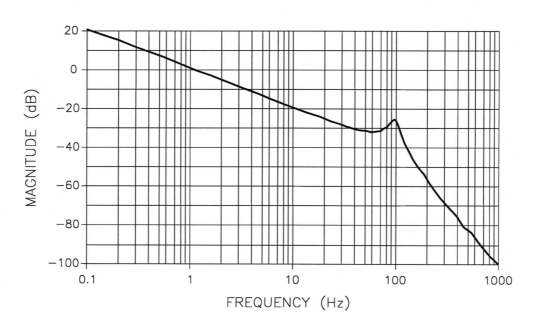

OPEN–LOOP MAGNITUDE TRANSFER FUNCTION

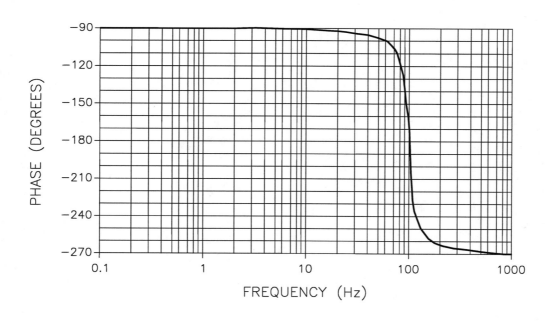

OPEN–LOOP PHASE TRANSFER FUNCTION

133. A single-phase transformer serves a 7.2-kW resistive load located 200 ft from the transformer. Each conductor from the transformer to the load has an impedance of $(0.78 + j0.052)$ Ω per 1,000 ft. If the voltage at the load terminals is 240 V, the voltage magnitude (V) at the **transformer secondary** is most nearly:

(A)   265
(B)   250
(C)   245
(D)   230

134. A 3-phase, 460-V, 25-hp induction motor draws 34 A at 0.75 lagging power factor from a 480-V source. The reactive power (kvar) required to correct the power factor to 0.90 lagging is most nearly:

(A)   −2.8
(B)   −4.9
(C)   −8.4
(D)   −14.6

135. The 3-phase load at a bus is 8,000 kW at 0.80 lagging power factor. The total reactive power (kvar) supplied by a 3-phase capacitor bank that will increase the power factor to 0.95 lagging is most nearly:

(A)   6,000
(B)   3,400
(C)   2,900
(D)   2,600

136. The diagram below represents the Thevenin equivalent of a single-phase distribution system. A fault occurs between point X and ground. R_F represents the fault resistance. The current I_F is 3,600 A when R_F is 0 Ω. If R_F is changed to 1.0 Ω, the current I_F (amperes) is most nearly:

(A) 2,000
(B) 2,400
(C) 3,200
(D) 4,600

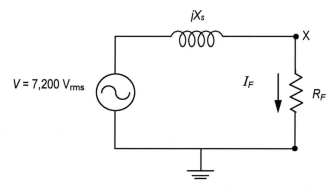

137. The equipment-grounding conductor for an electrical distribution system within a building is the conductor that connects noncurrent-carrying parts of equipment and raceways to ground at the service equipment or the source of a separately derived system. The purpose of an equipment-grounding conductor is to provide a:

(A) low impedance to ground to limit voltage to ground on exposed conducting surfaces and help ensure ground fault clearing

(B) path to ground for all harmonic currents

(C) path to ground for triplen harmonics

(D) low impedance to ground to limit voltage to ground on exposed conducting surfaces, help ensure ground fault clearing, and provide a path to ground for triplen harmonics

138. The one-line diagram below represents three single-phase transformers connected in a delta-wye arrangement and rated for the line-to-line voltages indicated as shown below. The turns ratio of individual single-phase transformers is most nearly:

(A)   2.8
(B)   4.8
(C)   8.3
(D)   14.4

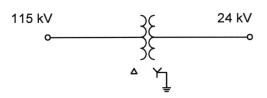

115 kV   24 kV

139. A 3-phase, 60-Hz induction motor has a nameplate speed of 1,600 rpm. This motor has how many poles?

(A)   2.0
(B)   4.0
(C)   4.5
(D)   6.0

                    GO ON TO THE NEXT PAGE

140. Consider the lossless ac transformer shown in the figure. The steady-state impedance (Ω) seen between Terminals H1-H2 is most nearly:

(A) $1\angle 30°$
(B) $8\angle -30°$
(C) $16\angle -150°$
(D) $16\angle 30°$

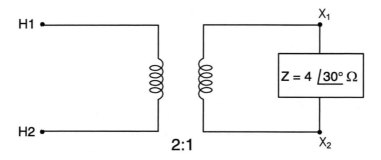

ELECTRICAL AND COMPUTER ENGINEERING

DEPTH AFTERNOON SAMPLE QUESTIONS

COMPUTERS
AFTERNOON SAMPLE QUESTIONS

501. The EIA RS-232C specification limits the maximum length (m) of the DTE cable to most nearly:

(A) 5
(B) 15
(C) 30
(D) 50

502. Consider the data byte $(CA)_{16}$ being transmitted serially and asynchronously using standard EIA RS-232C levels in the following signal diagram.

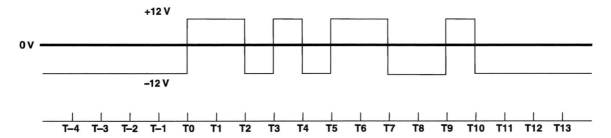

The bit group format in this transmission is:

(A) 1 start, 8 data, 1 parity, 1 stop with even parity
(B) 1 start, 8 data, 1 parity, 1 stop with odd parity
(C) 1 start, 7 data, 1 parity, 1 stop with even parity
(D) 1 start, 7 data, 1 parity, 1 stop with odd parity

503. A 16-bit floating point word has a format as shown in the following table.

Byte 0								Byte 1							
7	6	5	4	3	2	1	0	7	6	5	4	3	2	1	0
S	E							M							
	MSb						LSb	MSb							LSb

In the table, S is a sign bit with S = 1 indicating a value less than zero; E is a 7-bit excess-64 exponent of a power of 2; and M is an 8-bit mantissa that is a fractional part of unity. MSb and LSb denote the most significant and least significant bits of each byte in the word. A 16-bit word corresponding to $(4000)_{16}$ represents a numeric value that is most nearly:

(A)　0.0
(B)　0.5
(C)　1.0
(D)　2.0

COMPUTERS AFTERNOON SAMPLE QUESTIONS

504. An 8-bit microcontroller has an instruction set architecture as indicated in the table. The microcontroller uses a 3-bit instruction word with the following eight instructions:

Opcode	Mnemonic	Function	Description
000	JMP	PC⇐ACC	Jump to location in ACC
001	INC	ACC⇐ACC+1	Increment accumulator
010	ADD	ACC⇐ACC+B	Add B port data to accumulator
011	OUT	C⇐ACC	Output data to C port latch
100	CLR	ACC⇐0	Zero the accumulator
101	NOT	ACC⇐~ACC	Ones complement the accumulator
110	SHR	ACC[6:0]⇐ACC[7:1] ACC[7]⇐0	Shift accumulator 1 bit to the right
111	SHL	ACC[7:1]⇐ACC[6:0] ACC[0]⇐0	Shift accumulator 1 bit to the left

Make no assumptions concerning the initial content of the registers in this question. Assume that all data values are unsigned 8-bit numbers. Assuming that overflow does not occur, the proper code sequence to multiply the contents of the accumulator by 8 is:

- (A) SHL, SHL, SHL
- (B) SHR, SHR, SHR
- (C) CLR, INC, SHL, SHL, SHL
- (D) ADD, ADD, ADD, ADD, ADD, ADD, ADD, ADD

GO ON TO THE NEXT PAGE

505. The microprocessor-based data acquisition system shown below samples an analog signal, processes it digitally, and then converts the processed signal back to analog form. The system uses data converter circuits with the following characteristics:

A/D: 4-bit offset binary, unity gain, bipolar –5 to +5 V; –5 V corresponds to $(0000)_2$

D/A: 4-bit offset binary, unity gain, unipolar, +10 V; 0 V corresponds to $(0000)_2$

For an input voltage of +2.0 V, the A/D converter output code is most nearly:

(A) $(0011)_2$
(B) $(0100)_2$
(C) $(1011)_2$
(D) $(1101)_2$

506. A microcontroller system with an A/D converter is shown in the figure below. Assume that A15 and D7 are the most significant bits of the address and data buses.

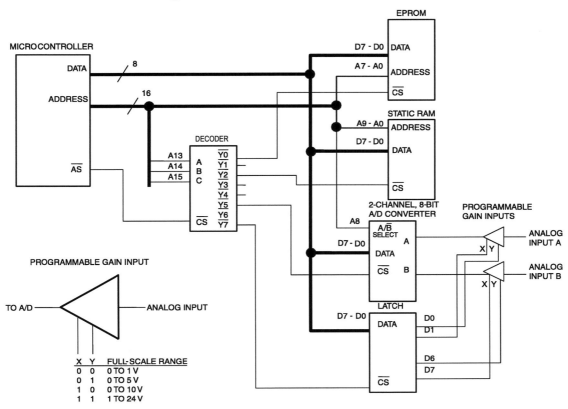

To set the programmable gain of analog input A for a full-scale input of 0 to 10 V, and input B for a full-scale input of 0 to 5 V, you must:

(A)   send the data $(01)_{16}$ to address location $(C000)_{16}$

(B)   send the data $(02)_{16}$ to address location $(DFFF)_{16}$

(C)   send the data $(83)_{16}$ to address location $(FFFF)_{16}$

(D)   send the data $(42)_{16}$ to address location $(FFFF)_{16}$

507. The propagation delay from CLK to OUT can be represented as the propagation delay through which gates?

(A) $\max (U_1, U_2) + U_4$
(B) $U_1 + U_2 + U_3 + U_4$
(C) $U_1 + U_2 + U_4$
(D) $U_1 + U_2$

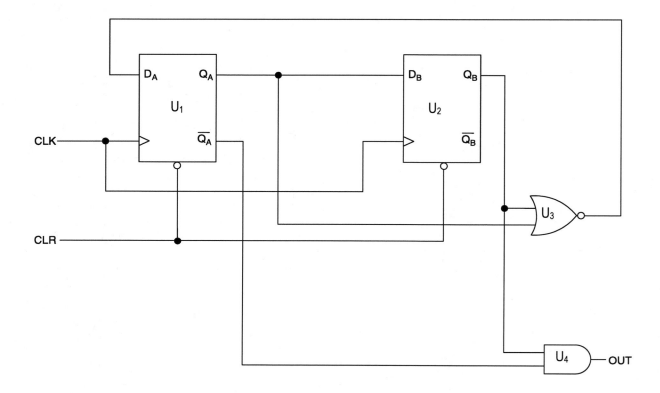

508. A Branch Control Logic Block (BCLB) is used within an Arithmetic Logic Unit (ALU) to generate the condition codes necessary for BRANCH ON CONDITION instruction codes. An overview of the proposed system is illustrated in the following diagram. The inputs to the BCLB system include two 4-bit words on the A and B data bus, with corresponding output control signals as follows:

- G [A>B] (A data bus value is greater than the B data bus value).
- E [A=B] (A data bus value is equal to the B data bus value).
- L [A<B] (A data bus value is less than the B data bus value).

Each BCLB element P_i has GT/LT/A/B inputs and G/L outputs that correspond to the overall system relations described below:

- GT - GREATER THAN input [A>B for previous stage]
- LT - LESS THAN input [A<B for previous stage]
- A - A data bus bit input
- B - B data bus bit input

All data values are to be interpreted as *unsigned* quantities, with Bit 3 being the MSb and Bit 0 the LSb. All control signals and data bus values are assumed to have positive logic interpretation.

The logic for output G of subelement P_i is most nearly:

(A) $G = GT + \overline{LT} \cdot A \cdot \overline{B}$

(B) $G = GT \cdot (A \cdot B + \overline{A} \cdot \overline{B})$

(C) $G = GT \cdot A \cdot \overline{B}$

(D) $G = GT \cdot \overline{LT}$

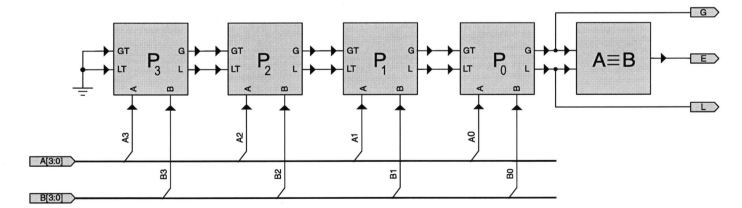

509. A CMOS clocked logic circuit is shown in the figure where Z is weakly pulled high during the PRECHARGE cycle of the CLK signal (CLK low) by the P-channel MOSFET M1. The logic value at Z is subsequently sampled during the EVALUATE cycle of the CLK signal (CLK high). Assume that the PRECHARGE cycle has been previously activated and that the CLK signal is currently at a logic high level. Assume positive logic.

The output Z has what logic function with respect to the A and B inputs?

(A) NOR
(B) OR
(C) NAND
(D) AND

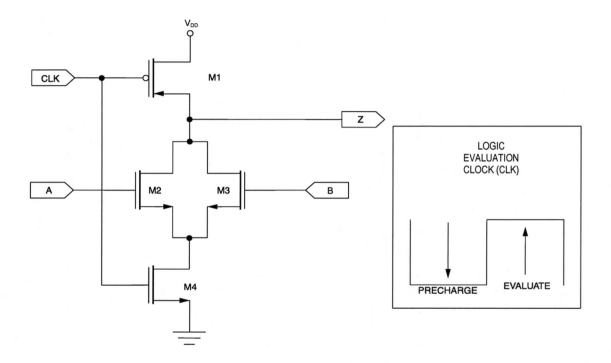

510. Consider the CMOS logic gate illustrated below. Digital inputs A and B represent logic inputs, and C is a digital control input.

When C = 0, what logic function is generated by the circuit at node Y?

(A) OR
(B) AND
(C) NOR
(D) XOR

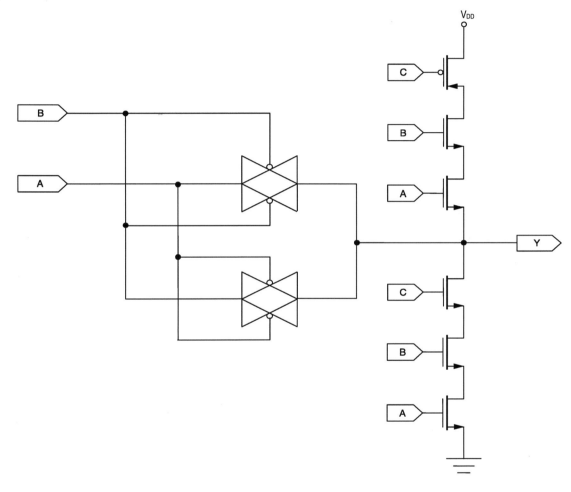

511. An ASCII character stream is being transmitted using the EIA RS-232C standard. The data format for transmission is 1 start bit, 8 data bits, an odd parity bit, and 1 stop bit. If the data transmission rate is 9,600 baud, the time (microseconds) required to transmit a single character frame is most nearly:

(A)   104
(B)   833
(C)   1,042
(D)   1,146

512. Input X is an asynchronous input to the synchronous state machine described by the state diagram below. Considering the possibility that input X may change at any time with respect to the state machine clock, the possible next states for State 01 are:

(A)   01, 10
(B)   01, 10, 00
(C)   00, 01, 10, 11
(D)   10, 00

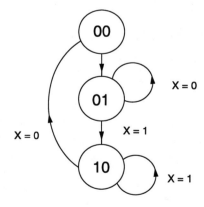

GO ON TO THE NEXT PAGE

513. The following circuit is used to synchronize an input from a slower system. Assume that the D flip-flops are positive edge triggered with zero second setup time.

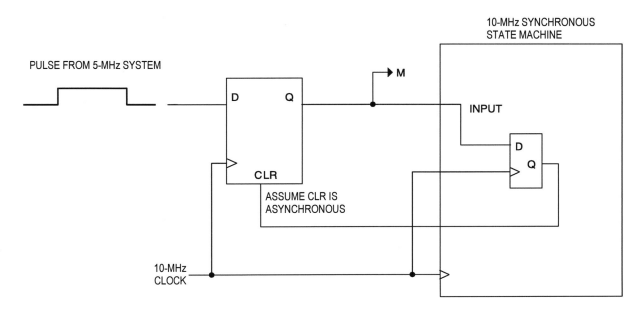

The most appropriate timing diagram describing the behavior of an input is:

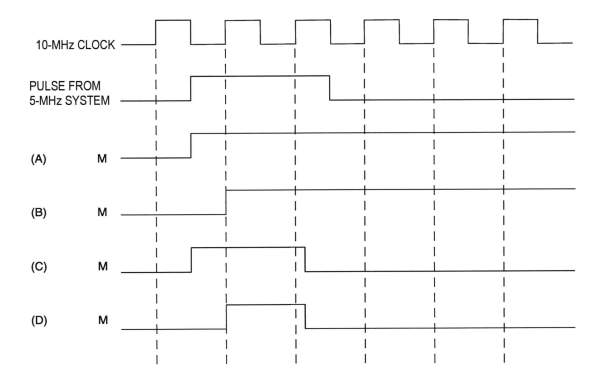

GO ON TO THE NEXT PAGE

514. For the system shown assume the flip-flops are positive edge triggered with a setup time of zero seconds.

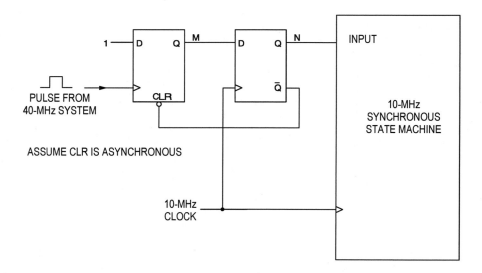

The most appropriate timing diagram is:

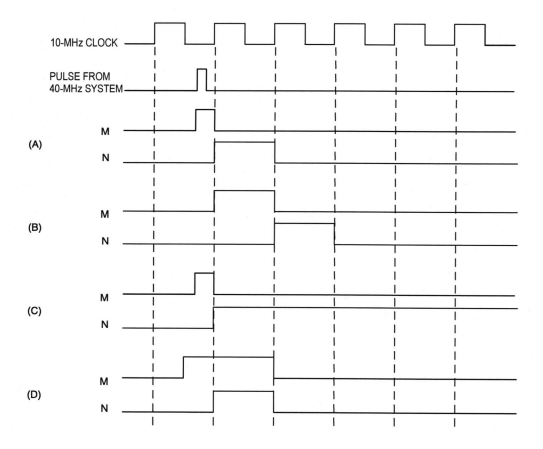

515. A vending machine accepts nickels, dimes, and quarters and dispenses candy only when a total of 25¢ or more has been reached. Symbols N, D, Q indicate a nickel input, a dime input, and a quarter input respectively, and the symbol C denotes a candy output. A state transition diagram that represents such a machine is given by:

(A)

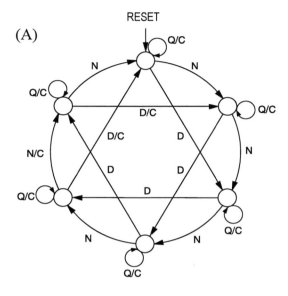

(B)

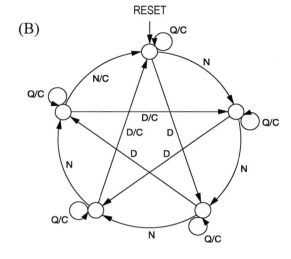

(C)

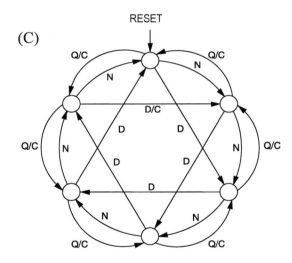

(D)

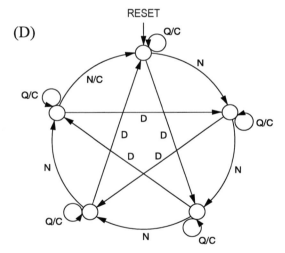

GO ON TO THE NEXT PAGE

516. The synchronous digital circuit below has input CLK and output Q_2. Assume positive logic. If the current state of the synchronous circuit is $Q_1Q_2Q_3 = 100$, the next state will be:

(A) 010
(B) 011
(C) 100
(D) 110

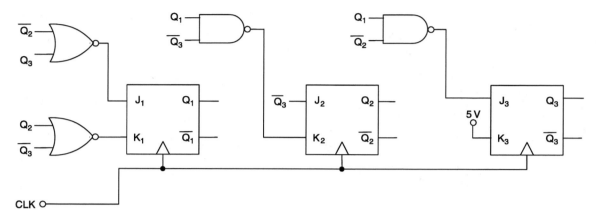

517. A synchronous counter circuit is implemented using a 3-bit Linear Feedback Shift Register (LFSR) as shown below. During the rising edge of each clock pulse, the LFSR shifts the register contents from the D_i position to the D_{i-1} position.

Assume that the LFSR is initialized with $D_2D_1D_0 = 001$. The number of states the circuit cycles through is most nearly:

(A) 2
(B) 3
(C) 7
(D) 8

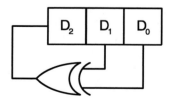

GO ON TO THE NEXT PAGE

518. A CPU contains a control unit that is designed with one-hot encoding in an attempt to maximize clock speed. This synchronous circuit cycles through 16 distinct states and produces 32 control signals. If the circuit is based on D flip-flop cells, the number of cells required is most nearly:

(A)   4
(B)   5
(C)   16
(D)   32

519. The block diagram in the figure below indicates a simple circuit that uses an arithmetic logic unit (ALU), accumulator, data registers, and an 8-bit instruction register (IR) to perform various computing functions. MSb and LSb denote the most significant and least significant bit respectively.

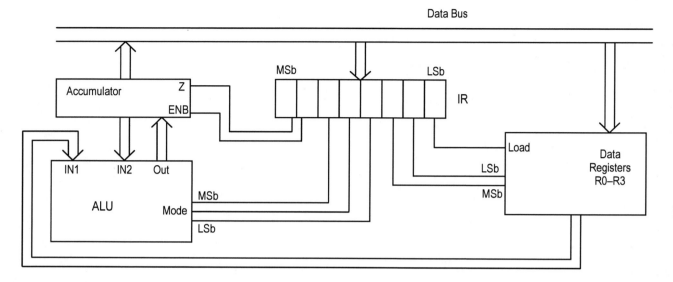

The ALU can perform the following functions based on the Mode input.

Mode MSb		LSb	Output
0	0	0	ZERO (output is zero)
0	0	1	IN1
0	1	0	IN2
0	1	1	NOT IN1 (output is the bitwise complement of IN1)
1	0	0	IN1 + IN2
1	0	1	IN1 − IN2
1	1	0	IN1 AND IN2
1	1	1	IN1 OR IN2

The output of the accumulator is always present at IN2 of the ALU. If the Z input to the accumulator is set to logic low, the output of the accumulator attached to the data bus is set to a high-impedance state. If the Enable input (ENB) is set high, then the accumulator will latch the output of the ALU on the next clock cycle.

Two bits are used to select which data register $[(00)_2 = R0, (01)_2 = R1, (10)_2 = R2, (11)_2 = R3]$ to input or output. If the load input is set high, then the register will input the value from the data bus on the next clock cycle.

GO ON TO THE NEXT PAGE

COMPUTERS AFTERNOON SAMPLE QUESTIONS

Assume the initial states of the accumulator and registers (in hex) are as follows:

Accumulator = $(FF)_{16}$
R0 = $(01)_{16}$
R1 = $(04)_{16}$
R2 = $(80)_{16}$
R3 = $(FF)_{16}$

Executing Instruction $(01110100)_2$ causes the accumulator to become most nearly:

(A) $(01)_{16}$
(B) $(04)_{16}$
(C) $(80)_{16}$
(D) $(FF)_{16}$

GO ON TO THE NEXT PAGE

520. For the computer whose machine instruction format and instruction set are shown, assume the instruction register contains $(AB6)_{16}$ after completion of an instruction fetch cycle. The "I" bit (when set) indicates an indirect address is present in bits 6:0. Note that "EA" means "effective address," the address of the operand. The mnemonic of the machine instruction is:

- (A) ldr
- (B) add
- (C) cad
- (D) iac

Opcode (binary)	Mnemonic	Description
0000	nop	No operation
0001	lac	Load AC
0010	ldr	Load DR
0011	sac	Store AC
0100	sdr	Store DR
0101	add	ADD AC + M [EA]
0110	and	AND AC, M [EA]
0111	xor	XOR AC, M [EA]
1000	not	Invert AC
1001	cda	Copy DR into AC
1010	cad	Copy AC into DR
1011	iac	Increment AC
1100	jmp	Jump to address
1101	skz	Skip next instr. If AC = 0
1110–1111	—	Unused

MACHINE-INSTRUCTION FORMAT

11	10 7	6 0
I	OPCODE	ADDRESS

GO ON TO THE NEXT PAGE

521. The machine instruction format illustrated uses one memory word per machine instruction. The maximum size (bits) of the user memory is most nearly:

(A) 2^7

(B) 12×2^7

(C) 2^{12}

(D) $2^{12} \times 2^7$

MACHINE-INSTRUCTION FORMAT

11	10	7 6		0
I	OPCODE		ADDRESS	

GO ON TO THE NEXT PAGE

COMPUTERS AFTERNOON SAMPLE QUESTIONS

522. Given the diagram shown, which of the following addresses will access Channel B of the A/D? Assume A15 and D7 are the most significant bits of the address and data buses.

(A) $(A100)_{16}$
(B) $(4000)_{16}$
(C) $(5000)_{16}$
(D) $(A000)_{16}$

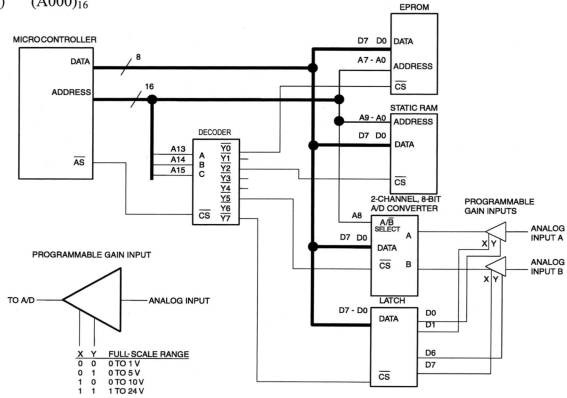

523. A four-input (variables = PABC) combinational circuit that serves as an interface between components of a special-purpose digital system is to be designed. The most significant bit P is a parity bit, and the circuit output $\overline{Z}$ is to be a 0 (indicating error) if the parity is *even*. An input combination PABC for which the output $\overline{Z}$ should be 0 (low) is:

(A) $(0010)_2$
(B) $(0100)_2$
(C) $(0101)_2$
(D) $(0111)_2$

524. An 8-bit microcontroller has a 3-bit instruction word with the following eight instructions:

Opcode	Mnemonic	Function	Description
000	JMP	PC⇐ACC	Jump to location in ACC
001	INC	ACC⇐ACC+1	Increment accumulator
010	ADD	ACC⇐ACC+B	Add B port data to accumulator
011	OUT	C⇐ACC	Output data to C port latch
100	CLR	ACC⇐0	Zero the accumulator
101	NOT	ACC⇐~ACC	Ones complement the accumulator
110	SHR	ACC[6:0]⇐ACC[7:1] ACC[7]⇐0	Shift accumulator 1 bit to the right
111	SHL	ACC[7:1]⇐ACC[6:0] ACC[0]⇐0	Shift accumulator 1 bit to the left

Make no assumptions concerning the initial contents of registers in this question. Assume that all data values are unsigned 8-bit numbers. All instructions have equal execution times. A continuous square wave with a 50% duty cycle is to be generated at C[0] the LSb of output port C. The sequence of instructions that will generate the square wave after the first six in the sequence have executed is:

(A)   NOT, OUT, CLR, SHR, OUT, JMP
(B)   CLR, ADD, OUT, ADD, CLR, JMP
(C)   CLR, OUT, NOT, OUT, NOT, JMP
(D)   CLR, NOT, OUT, SHL, OUT, JMP

525. Consider the performance of an arbitrary CPU running a particular application program instruction mix. The table below provides the instruction frequencies and the time distributions for the CPU application. The fraction of time the program spends doing CALLs, RETs, PUSHes, and POPs is most nearly:

(A)   5%
(B)   6%
(C)   11%
(D)   21%

Instruction Frequencies and Time Distributions			
Instruction	**Clocks Per Instruction (CPI)**	**Instruction Frequency (%)**	**Time Distribution (%)**
CONTROL		24	24
CALL, CALLF	16	4	6
RET, RETF	20	4	5
LOOP (iterate loop)	12	4	3
Conditional jumps (JNE, etc.)	24	10	7
JMP	12	2	3
ARITHMETIC, LOGICAL		25	15
CMP (compare value)	4	7	6
TEST (test value)	4	1	1
SAL, SHR, RCR	2	5	3
OR, XOR	2	3	1
ADD	2	3	1
SUB	2	2	1
INC, DEC	2	3	1
CBW (convert byte to word)	2	1	1
DATA TRANSFER		42	35
MOV	16	27	21
LES (load address)	10	3	4
PUSH (data or CPU status flags)	14	7	7
POP (data or CPU status flags)	12	5	3
TOTALS		91	74

526. Consider the cyclic redundancy check (CRC) generator in the following block diagram. Boxes represent D flip-flops clocked from a common clock. The input data stream is stable before a clock edge and is injected synchronously with the rising edge of the clock.

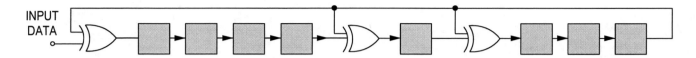

The CRC generator polynomial is:

(A) $X^8 + X^4 + X^2 + 1$
(B) $X^8 + X^4 + X^3 + 1$
(C) $X^8 + X^5 + X^4 + 1$
(D) $X^8 + X^5 + X^4 + 8$

527. The parity-check matrix H for a 7-bit even-parity Hamming Code is shown below. The format of the codeword is $D_6D_5D_4D_3D_2D_1D_0$ and is composed of the information bits $I_3I_2I_1I_0$ and the parity bits $P_2P_1\,P_0$ as follows:

MSb   LSb

D_6	D_5	D_4	D_3	D_2	D_1	D_0
I_3	I_2	I_1	P_2	I_0	P_1	P_0

$$H = \begin{bmatrix} 1 & 1 & 1 & 1 & 0 & 0 & 0 \\ 1 & 1 & 0 & 0 & 1 & 1 & 0 \\ 1 & 0 & 1 & 0 & 1 & 0 & 1 \end{bmatrix}$$

Given information bits $(1\ 1\ 1\ 0)_2$, what is the Hamming Code word?

(A) $(1\ 1\ 1\ 0\ 1\ 0\ 0)_2$
(B) $(1\ 1\ 1\ 1\ 0\ 0\ 0)_2$
(C) $(0\ 0\ 0\ 1\ 1\ 1\ 1)_2$
(D) $(1\ 1\ 1\ 0\ 0\ 1\ 1)_2$

528. A structured data-type made up of a finite collection of fields that are not necessarily homogenous elements is called a:

(A)   binary tree
(B)   record
(C)   stack
(D)   queue

529. A function EUCLID(m,n) is constructed to return the greatest common divisor (GCD) of integers m and n using the Euclidean algorithm. The Euclidean algorithm has the following form:

E1: Divide m by n and let r be the remainder
E2: If r = 0, the algorithm terminates; n is the answer
E3: Set m = n, n = r, and proceed to E1.

Executing the procedure call EUCLID (440, 200) requires how many total levels of recursion?

(A)   1
(B)   2
(C)   3
(D)   4

530. Software program errors or "bugs" cause a program not to perform its intended function. Listed below are several types of program bugs. Among those listed below, the error that generally requires the **LEAST** amount of resources to fix is:

(A)   A program that has been written to the wrong specification, i.e., the program runs but does not accomplish its intended function
(B)   Compile-time errors
(C)   Run-time errors
(D)   An error caused by incorrect user input

531. Software verification activities begin when software specifications are developed. At this point, the overall testing goals and approach are formulated. Listed below are steps in the software verification process.

    I. Test cases are developed and test data to support them are generated.

    II. Unit testing, or module-level testing, is done with stubs and drivers used for support.

    III. Interface tests are performed to test the interoperability of modules.

    IV. Tests performed earlier are rerun as regression tests to make sure that any corrections implemented during the software verification process have not introduced new problems.

Assuming top-down testing techniques are applied, list in order the steps that must be accomplished for software verification.

    (A)   I, II, III, IV
    (B)   I, III, II, IV
    (C)   II, III, I, IV
    (D)   III, II, I, IV

532. Simulation is a technique in which models of objects and events are used to predict the outcome of a physical event. Which of the following is a **FALSE** statement concerning simulation techniques?

    (A)   A model can be thought of as a series of statements or rules that describe the behavior of a real-world system.

    (B)   In the simulation, each object in the real-world system is usually represented by a data object. Actions in the real world are represented as operations on the data object.

    (C)   If the simulation does not accurately model the real-world event, the real-world event is studied in great detail to make it match the simulation.

    (D)   A random-number generator is often used to introduce random events to the simulation.

COMPUTERS AFTERNOON SAMPLE QUESTIONS

533. In a CPU, instruction pipelining is a technique to decrease the clock cycles per instruction (CPI) for a processor. The **FALSE** statement concerning a pipelined system is:

    (A)   The data path must be separated into consecutive stages.

    (B)   The pipeline stage with the shortest latency determines the maximum clock frequency.

    (C)   Stalls prevent the pipeline from achieving a CPI of 1.

    (D)   The pipeline improves performance by increasing throughput, as opposed to decreasing the execution time of an individual instruction.

534. Of the following statements, the most complete description of software documentation is:

    (A)   Software documentation is the comments used throughout the source code to make the program understandable.

    (B)   Software documentation is the written information that is outside the body of the source code. In addition to the detailed software specifications, it may include the history of the program's development and subsequent modifications.

    (C)   Software documentation includes comments, self-documenting code, and program formatting for readability.

    (D)   Documentation includes the written descriptions, specifications, design, and actual source code of a program. It includes both external and internal documentation.

GO ON TO THE NEXT PAGE

535. Of the following statements concerning the software design technique of successive refinement, the one that is **NOT** true is:

(A)   The successive refinement technique takes a divide-and-conquer approach. That is, the problem is broken into several large tasks. Each of these tasks is then divided into sections, the sections are subdivided, and so on.

(B)   Details of implementation are dealt with at the highest levels.

(C)   Details are deferred as long as possible as we go from a general to a specific solution.

(D)   Successive refinement is a key element of the top-down design approach.

536. $A \times (B + C \times D)$ is an expression in infix form. The equivalent postfix expression is most nearly:

(A) $\times A + B \times C\ D$
(B) $A\ B\ C\ D \times + \times$
(C) $A\ B \times C\ D + \times$
(D) $A\ B\ C\ D \times \times +$

537. A computer network is modeled as a graph where the vertices represent computers and the edges represent network connectivity. The **diameter** is a convenient metric used to estimate throughput by the network engineer. Assuming each edge has unity weight, the diameter of the network in the figure is:

(A)　1
(B)　2
(C)　3
(D)　4

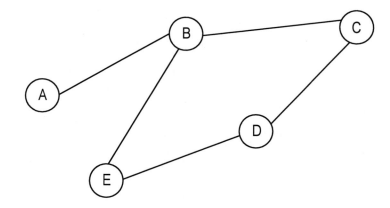

538. A business uses workstations that are networked together using a TCP/IP protocol stack. The TCP layer corresponds most closely to the following ISO/OSI model layer:

(A)　Layer 1, the physical layer
(B)　Layer 2, the data link layer
(C)　Layer 3, the network layer
(D)　Layer 4, the transport layer

539. In a 7-layer open system interconnection (OSI) system, the physical, data link, and network layers, respectively, correspond to the following layers:

(A) 3, 5, 7
(B) 5, 6, 7
(C) 1, 2, 3
(D) 1, 3, 5

540. A cable television transmission line with a characteristic impedance of 75 Ω carries a broadband signal from a transmitter to a receiver. It is decided to monitor the signal by bridging (connecting in parallel) a test set across the line at a point remote (so that the line loss exceeds 10 dB) from both the transmitter and the receiver. The test set has a resistive input impedance of 50 Ω, with an additional resistor of 47 kΩ (with negligible capacitance at the operating frequency) in series with the test set.

While the 47-kΩ resistance and the test set are bridged across the transmission line, the impedance of the line as measured at the receiver, relative to the impedance seen prior to the connection of the set, will most nearly:

(A) change by a factor of 2
(B) change by a factor of 1.5
(C) remain essentially unchanged
(D) change by a factor of 2/3

ELECTRONICS, CONTROLS, AND COMMUNICATIONS
AFTERNOON SAMPLE QUESTIONS

501. This question investigates the performance of the temperature measurement circuit illustrated in the figure below.

The temperature sensor is linear, with the following open-circuit voltages:
$V_S = 0$ mV at 0°C
$V_S = -100$ mV at 100°C

The operational amplifier (op amp) is ideal.

At 100°C, the maximum output voltage at V_o (V) due to R_2 tolerance is most nearly:

(A) 2.5
(B) 5.2
(C) 10.1
(D) 15.3

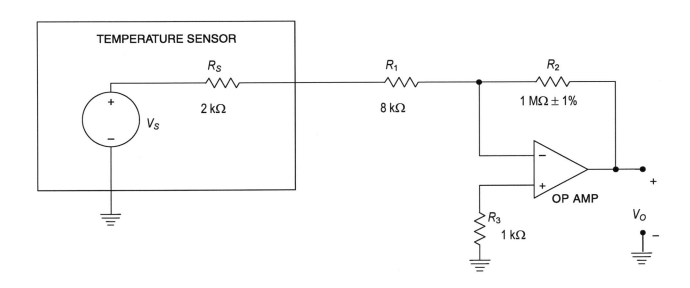

502. Which of the following regression models is **NOT** considered to be a Linear Regression Model to predict Y in terms of X?

(A) $\hat{Y} = b_0 + b_1 X_1$

(B) $\hat{Y} = b_0 + b_1 X_1 + b_2 X_2^2$

(C) $\hat{Y} = b_0 + b_1 X_1 + b_2 X_2 + b_{12} X_1 X_2$

(D) $\hat{Y} = b_0 + b_1 X_1^{b_2}$

503. A 5-V microcontroller has an RS-232 interface connected to a piece of data communications equipment (DCE). The RS-232 line drivers in both the microcontroller and the DCE are powered by ± 12 V. The voltage (V) measured at the Receive Data (RXD) wire of the interface with respect to the Signal Ground (SG) wire during idle data transmission conditions is most nearly:

(A) +12
(B) +5
(C) 0
(D) −12

504. An engineer chooses open-collector output TTL technology in order to provide sufficient current to drive a device. The circuit shown below is implemented with a 7405 package and corresponds to the following logic function (assume that a logic-0 corresponds to 0 V):

(A) AND
(B) NAND
(C) OR
(D) NOR

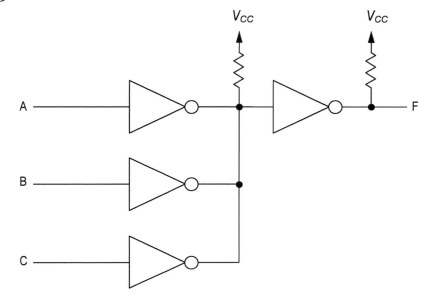

505. The transfer function, $A_v(s)$, of an active filter is given by

$$A_v(s) = \frac{-10^6 s}{s^2 + 10^5 s + 10^{12}}$$

The 3-dB bandwidth (rad/s) is most nearly:

(A) 1×10^5
(B) 2×10^5
(C) 5×10^5
(D) 1×10^6

GO ON TO THE NEXT PAGE

506. An electrochemical cell exhibits the nonlinear V-I characteristic shown in **Figure 1** and is used in the circuit shown in **Figure 2.**

The output voltage V_o (V) in **Figure 2** is most nearly:

(A) 2.0
(B) 4.0
(C) 6.0
(D) 8.0

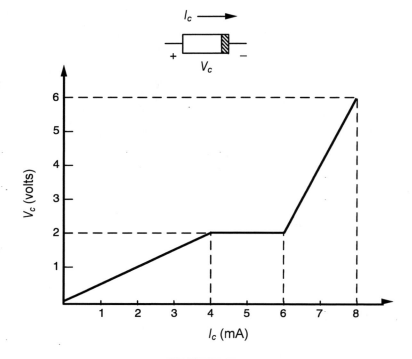

FIGURE 1

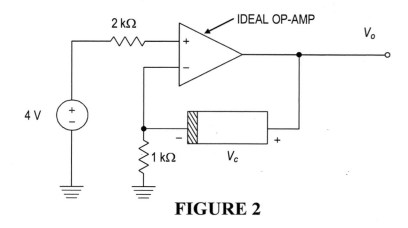

FIGURE 2

GO ON TO THE NEXT PAGE

507. This question investigates the power supply considerations of IC package designs. The package configuration for a digital latch is illustrated in **Figure 1**. **Figure 2** illustrates the circuit model for the package pin-to-pad inductance and capacitance. The table below provides the L and C model values for each pin.

Pin/Pad Model Values					
Pin	C (pF)	L (nH)	Pin	C (pF)	L (nH)
1	1.49	10.97	20	1.49	10.97
2	1.29	8.63	19	1.29	8.63
3	0.80	6.01	18	0.80	6.01
4	0.72	4.02	17	0.72	4.02
5	0.53	3.00	16	0.53	3.00
6	0.53	3.00	15	0.53	3.00
7	0.72	4.02	14	0.72	4.02
8	0.80	6.01	13	0.80	6.01
9	1.29	8.63	12	1.29	8.63
10	1.49	10.97	11	1.49	10.97

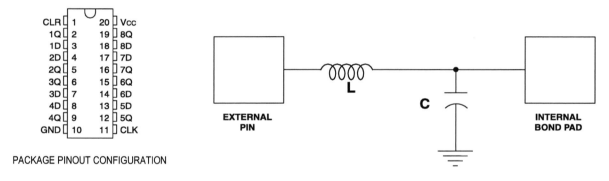

PACKAGE PINOUT CONFIGURATION

FIGURE 1

PIN TO PAD PACKAGING MODEL

FIGURE 2

The effective power supply inductance (nH) is most nearly:

(A)   2
(B)   5
(C)   10
(D)   20

508. The collector current I_{C3} is most nearly:

(A) $\dfrac{20 - V_{BE3}}{R_{C3}}$

(B) $\dfrac{(20 - V_{BE3})}{(1 + 2/\beta)R_{C3}}$

(C) $\dfrac{(20 - V_{BE3})}{(1 + 2\beta)R_{C3}}$

(D) $\dfrac{(20 - V_{BE3})(1 + 2/\beta)}{R_{C3}}$

$\beta = \beta_3 = \beta_4$

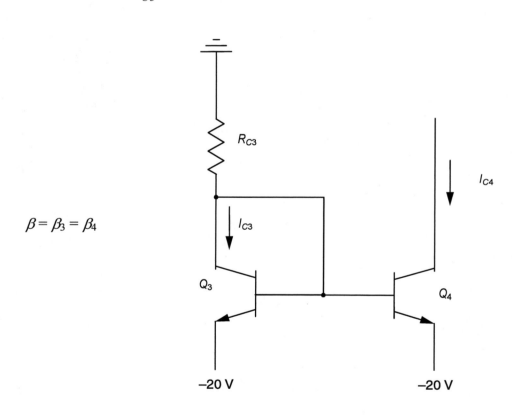

509. The radiation pattern of a wire of length L (centered at $x = 0$, $y = 0$, and $z = 0$) is shown in the figure below. The pattern is normalized such that the magnitude of radiation is one in the direction of maximum radiation. The wire is located along the y-axis of the radiation pattern as shown by the broader line. The wire has a time-varying current distribution along its length.

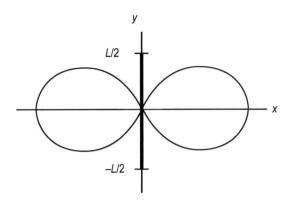

The current distribution that would give the above pattern is most nearly described by:

(A) $I_o \cos\left(\dfrac{\pi y}{L}\right)$

(B) $I_o \cos\left(\dfrac{3\pi y}{L}\right)$

(C) $I_o \cos\left(\dfrac{5\pi y}{L}\right)$

(D) $I_o \cos\left(\dfrac{7\pi y}{L}\right)$

510. A transmitter operating at 23.6 GHz has a power output of 20 mW. A receiver is located 8 miles from the transmitter.

The free space attenuation (dB) is most nearly:

(A) 142
(B) 124
(C) 119
(D) 115

511. A long transmission line has a loss of 3 dB at 900 MHz. If the far end of the line is accidentally short-circuited, what VSWR will be measured at the **input end**?

(A) 1:1
(B) 3:1
(C) 10:1
(D) ∞:1

GO ON TO THE NEXT PAGE

512. This question investigates the design of a series regulated power supply, shown in **Figure 1**, using a conventional 5-V series regulator. Assume the load (R_{LOAD}) is purely resistive and negligible current passes through a series regulator common ground connection.

Figure 2 illustrates the case temperature dissipation derating curve for the series regulator. The junction/case thermal resistance for the series regulator is provided as noted in the figure.

Assume V_{IN} = 14 V, R_{LOAD} = 5 Ω, with a heat sink applied to the case. The maximum permissible case temperature under these conditions is most nearly:

(A)   110°C
(B)   112°C
(C)   114°C
(D)   116°C

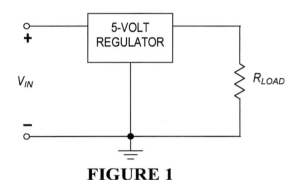

FIGURE 1

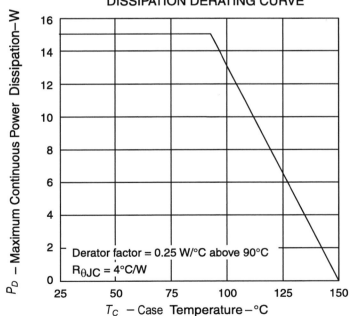

FIGURE 2

513. The circuit shown in the figure is for a simple 3-phase bridge inverter. It shows the power source (battery) and six switchable transistors and their bypass diodes. The goal is to provide a 3-phase waveform, in this case, square waves. The state of the transistors is indicated by the binary word (abcdef) using "0" to denote "off" or not conducting and "1" to denote "on" or conducting. For example, (000000) denotes that the whole system is off, and (010000) means that only transistor "b" is conducting.

Assume that phase "A" starts at "0" degrees, then at 90 degrees phase "AB" is a positive V, phase "BC" is zero, and phase "CA" is a negative V. What is the six-digit code for the transistors, (xxxxxx)?

(A)   (100001)
(B)   (100110)
(C)   (100101)
(D)   (101100)

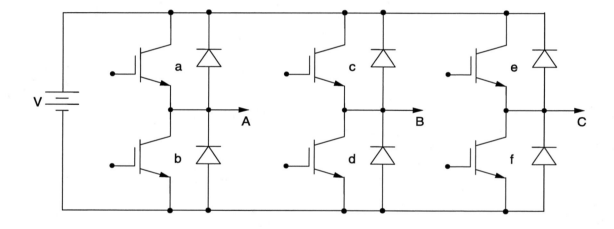

514. Consider the oscillator circuit shown below.

This circuit uses which resonant mode(s) of the crystal X1?

(A)   Series resonant
(B)   Parallel resonant
(C)   Series-parallel resonant
(D)   Not resonant

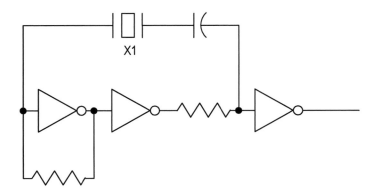

GO ON TO THE NEXT PAGE

515. The circuit contains a PNP bipolar junction transistor with $V_{EB(ON)} = 0.7$ V in the active and saturation modes, $V_{EC(SAT)} = 0.5$ V in the saturation mode, and common emitter current gain $\beta_F = 100$.

The magnitude of the transistor dc collector current (mA) is most nearly:

(A)   0.40
(B)   0.62
(C)   0.93
(D)   1.43

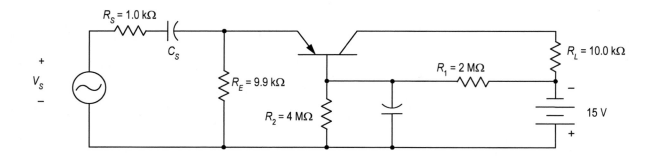

GO ON TO THE NEXT PAGE

516. Consider the following circuit with an N-Channel enhancement mode MOSFET:

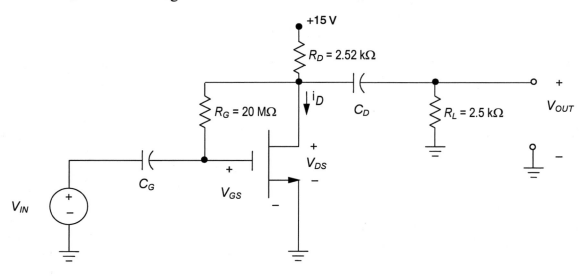

C_D and C_G are infinitely large.

The small signal model for the FET is:

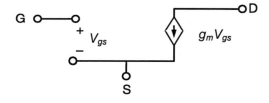

$\dfrac{V_{out}}{V_{in}}$ equals:

(A) $\dfrac{-g_m R_G}{1 - \dfrac{R_G}{R_D} - \dfrac{R_G}{R_L}}$

(B) $\dfrac{1 - g_m R_G}{1 - \dfrac{R_G}{R_D} - \dfrac{R_G}{R_L}}$

(C) $\dfrac{-g_m R_G}{1 + \dfrac{R_G}{R_D} + \dfrac{R_G}{R_L}}$

(D) $\dfrac{1 - g_m R_G}{1 + \dfrac{R_G}{R_D} + \dfrac{R_G}{R_L}}$

GO ON TO THE NEXT PAGE

517. Assume the op-amp in this circuit is ideal. In order that $V_{out} = K(V_1 - V_2)$, the following relationship between resistors must be true:

(A) $(R_1 + R_2) = (R_3 + R_4)$

(B) $(R_1 + R_3) = (R_2 + R_4)$

(C) $\dfrac{R_1}{R_2} = \dfrac{R_3}{R_4}$

(D) $\dfrac{R_1}{R_4} = \dfrac{R_3}{R_2}$

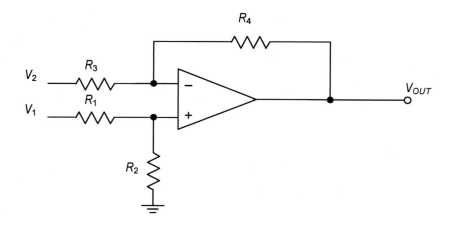

518. Consider the MOSFET degenerated current mirror shown below having input current I_X and output current I_Z. Assume the following conditions.

The MOSFETs have the following characteristics:

- Identical electrical characteristics
- Threshold voltages of 1 V
- Infinite drain output resistance (R_{DS})
- Operate within their saturated region of operation

All current sources are ideal.

In the saturation region, $I_D = K[V_{GS} - V_T]^2$ where $K = 50$ μA/V² for each MOSFET.

Assuming $I_Z = 10$ μA, and the voltage at node X is 5 V, the resistance (kΩ) of resistor R is most nearly:

(A) 52
(B) 118
(C) 237
(D) 355

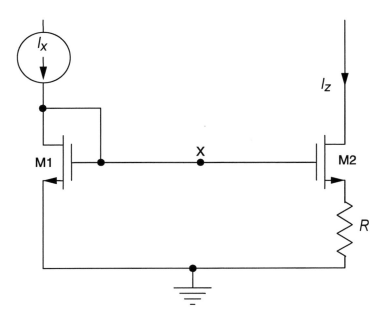

519. Given a pulse transfer function of:

$$\frac{1}{z^{-1}\left(1-0.7z^{-1}\right)}$$

The finite pole(s) in the z-plane is/are:

(A)   0 and 0.7
(B)   0 and 1/0.7
(C)   0.7
(D)   1/0.7

520. Consider the following closed-loop feedback system, where K can be varied from 0 to $+\infty$. K has been chosen so that the system is stable and $x(t)$ is a unit step function occurring at $t = 0$.

$y(t = 0)$ is most nearly:

(A)   0
(B)   16.7
(C) K
(D) $10K$

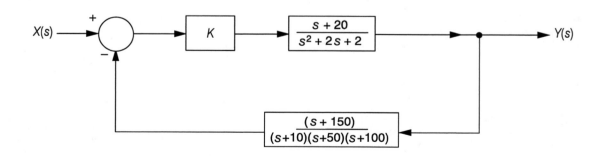

521.   Given the following block diagram:

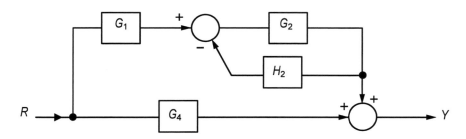

The transfer function $\dfrac{Y}{R}$ is:

(A) $\dfrac{G_1G_2}{1+G_2H_2} + G_4$

(B) $\dfrac{G_1G_2}{1-G_2H_2} + G_4$

(C) $G_1 + \dfrac{G_2}{1+G_2H_2} + G_4$

(D) $\dfrac{G_1G_2G_4}{1+G_2H_2}$

522. The control system given in the figure is to be formulated in state variable form. The generalized equations for the state variable formulation are:

$$\frac{dx(t)}{dt} = Ax(t) + Bu(t)$$ I.   State equation

$$y(t) = Cx(t)$$ II.   Output equation

The state variables are to be defined as $x(t) = x_1(t)$, $dx(t)/dt = x_2(t)$.

Let the state matrix A in equation I be defined as

$$\begin{bmatrix} 0 & 1 \\ -6 & -3 \end{bmatrix}$$

Using the A matrix as defined, determine the matrix C.

(A) $[1 \quad 4]$

(B) $\begin{bmatrix} 1 \\ 4 \end{bmatrix}$

(C) $[4 \quad 1]$

(D) $\begin{bmatrix} 4 \\ 1 \end{bmatrix}$

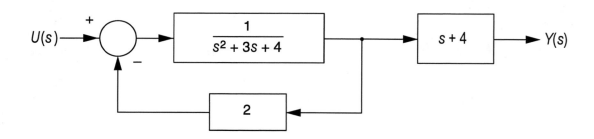

523. The type of controller used in the system below is:

(A)   lead
(B)   lag
(C)   integral
(D)   lead-lag

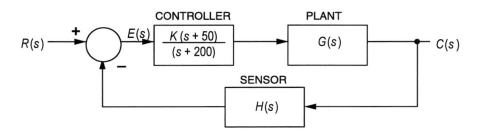

FEEDBACK CONTROL SYSTEM

524. A passive lumped-parameter RC network used as a lead-lag or lag-lead compensator in the feedback path is which type of controller?

(A)   PD
(B)   PI
(C)   PID
(D)   None of the above

525. The root locus of a control system is shown in the figure below. The plant has a transfer function of

$$G_p(s) = \frac{K}{s(s+1)(s+5)}$$

and the compensator, $H(s)$, is in the forward path of the closed-loop system. Let A be a constant.

The transfer function, $H(s)$, of the compensator is most nearly:

(A) $A\dfrac{s+1}{s+10}$

(B) $A\dfrac{s+10}{s+1}$

(C) $A\dfrac{s+5}{s+10}$

(D) $A\dfrac{s+1}{s+5}$

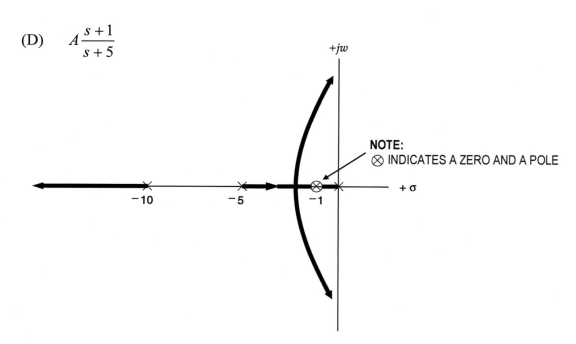

526. If the customer specifies an open-loop phase margin of 60°, the value of K that most nearly meets the specification is:

(A) 182

(B) 143

(C) 57

(D) 11

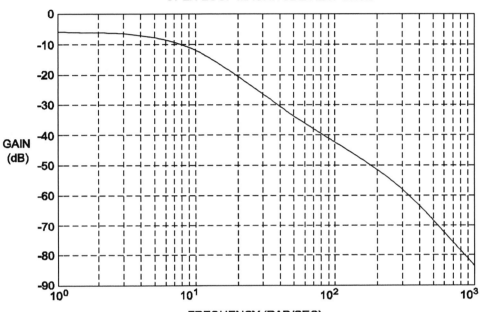

OPEN-LOOP MAGNITUDE RESPONSE

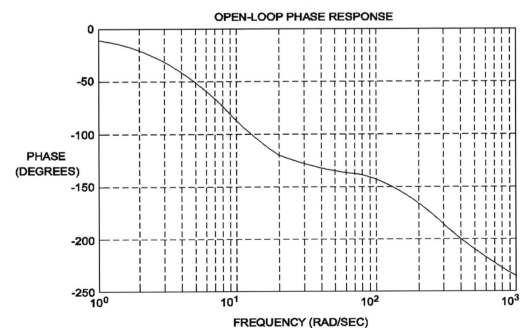

OPEN-LOOP PHASE RESPONSE

OPEN-LOOP BODE PLOT WITH $K=1$

GO ON TO THE NEXT PAGE

527.   **Figure 1** shows the magnitude (dB) of the open-loop transfer function of a feedback control system as a function of frequency, and **Figure 2** shows the phase of the open-loop transfer function as a function of frequency. Answer the following question based on these figures.

OPEN−LOOP MAGNITUDE TRANSFER FUNCTION

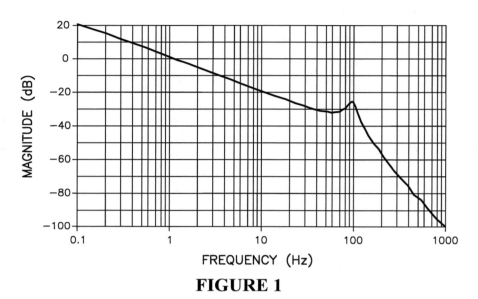

FIGURE 1

OPEN−LOOP PHASE TRANSFER FUNCTION

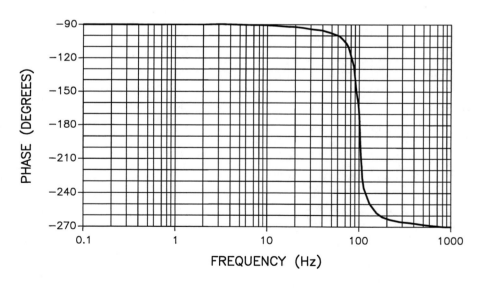

FIGURE 2

GO ON TO THE NEXT PAGE

ELECTRONICS, CONTROLS, AND COMMUNICATIONS
AFTERNOON SAMPLE QUESTIONS

The open-loop poles of the system are most like which of the following s-plane figures?

(A)

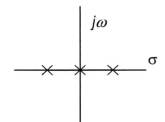

(B)

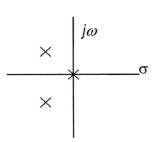

(C)

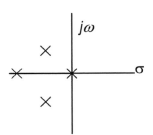

(D)

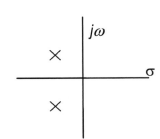

GO ON TO THE NEXT PAGE

528. The open-loop gain margin (dB) for the system whose magnitude and phase responses are shown below is most nearly:

(A) 6
(B) 32
(C) 43
(D) 56

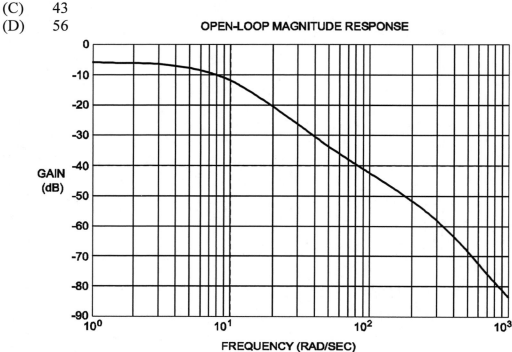

OPEN-LOOP MAGNITUDE RESPONSE

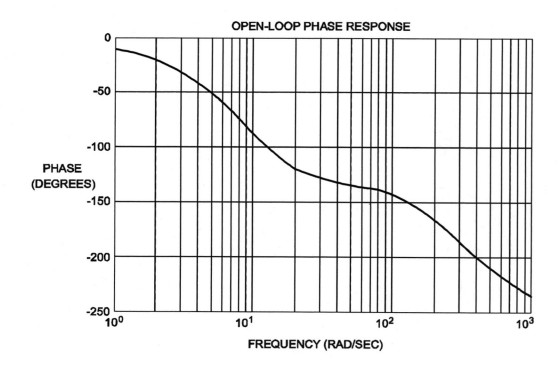

OPEN-LOOP PHASE RESPONSE

529. Consider the voltage waveform, $x(t)$, which is the output of an amplitude-modulated (AM) transmitter:

$$x(t) = 40 \cos (200{,}000\pi \, t) + 10 \cos (197{,}000\pi \, t) + 10 \cos (203{,}000\pi \, t)$$

The percentage of modulation of the AM waveform, $x(t)$, is most nearly:

(A) 25
(B) 33
(C) 50
(D) 75

530. An FM modulator has input $m(t) = 8 \cos (8\pi \, t)$. The peak frequency deviation of modulator output is 64 Hz. The modulation index is most nearly:

(A) 4
(B) 8
(C) 16
(D) 64

GO ON TO THE NEXT PAGE

531. An FM transmitter has an output waveform that can be written as

$$x(t) = A \cos [6.2 \times 10^{11}t + 10 \cos (10^7 t)].$$

The bandwidth (Hz) of the signal is most nearly:

(A) 1.59×10^6
(B) 1.0×10^7
(C) 1.59×10^7
(D) 3.5×10^7

532. The most common circuit used to demodulate an AM waveform is:

(A) a bandpass limiter
(B) an envelope detector
(C) a balanced discriminator
(D) a phase-locked loop

533. An FM transmitter has an output waveform that can be written as

$$x(t) = A \cos [6.2 \times 10^{11}t + 10 \cos (10^7 t)].$$

The peak instantaneous frequency deviation (Hz) from the carrier frequency is most nearly:

(A) 1.0×10^1
(B) 1.59×10^6
(C) 1.59×10^7
(D) 6.2×10^{11}

534. According to the Nyquist sampling theorem, a sufficient condition for there to be no aliasing in the sampled signal is that the highest frequency in the signal prior to sampling must be:

(A) less than half of the sampling frequency.
(B) greater than half of the sampling frequency.
(C) less than the sampling frequency.
(D) greater than the sampling frequency.

535. Consider the three-stage r-f amplifier shown below. AR1 refers to the input amplifier stage, AR2 to the second stage, and AR3 to the output stage.

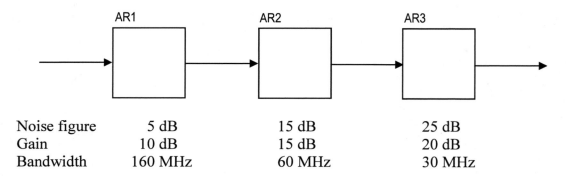

	AR1	AR2	AR3
Noise figure	5 dB	15 dB	25 dB
Gain	10 dB	15 dB	20 dB
Bandwidth	160 MHz	60 MHz	30 MHz

For this amplifier configuration, the total noise figure (dB) is most nearly:

(A)   5.0
(B)   6.6
(C)   8.6
(D)   12.3

536. An antenna with an effective aperture of 1 m^2 is connected to a receiver. In the absence of any signal, the local noise produces a received power of –40 dBm. If an incoming signal must produce a S/N of 10 dB, the RMS electric field intensity (mV/m) is most nearly:

(A)   0.37
(B)   1.0
(C)   7.1
(D)   19.5

537. It is desired to use a 50-Ω resistor as a low-level broadband noise generator for a matched, sensitive receiver with a 10-MHz bandwidth. The desired rms noise voltage across the resistor is 1.5 µV, when measured with a bandwidth of 10 MHz. The noise temperature (K) of the resistor must be most nearly:

(A)   110
(B)   290
(C)   330
(D)   400

538. A basic "Time-Domain Reflectometry" (TDR) system is used with an oscilloscope to display the waveform at the input end of a line driven by an alternating 0–1 square waveform signal in a non-return-to-zero (NRZ) digital system.

A 1,000-m cable is terminated with a short circuit at the far end. The first reflected wave appears 9.43 μs after the incident waveform. The actual velocity in the cable divided by the speed of light is most nearly:

(A)   0.35
(B)   0.71
(C)   1.0
(D)   1.41

539. Digital telecommunication equipment provides connection-oriented AND/OR connectionless service. Connection-oriented service is provided by:

(A)   ATM
(B)   ETHERNET
(C)   FDDI
(D)   SMDS

540. The termination of a transmission line is such that when 5 W propagate down the cable, 1 W is reflected. The absolute value of the reflection coefficient is most nearly:

(A)   0.167
(B)   0.200
(C)   0.447
(D)   0.833

POWER
AFTERNOON SAMPLE QUESTIONS

POWER AFTERNOON SAMPLE QUESTIONS

501. A SCADA system configuration requires maximum values (i.e., the actual value of current when the transducer is at full scale). The transducer has a full-scale ac current of 5 A. Using a current transformer with a turns ratio of 400:5, the maximum value of current (amperes) is most nearly:

  (A) 1,200
  (B) 800
  (C) 400
  (D) 300

502. A differential relay circuit connected to protect a delta-wye transformer using current transformers (CTs) must be balanced with properly connected auxiliary CTs to allow for the 30° phase shift and the turns ratio. If this is not accomplished, which of the following is most correct?

  (A) The circuit could cause the relay to be picked up under non-fault conditions.

  (B) The American national standard for designating terminals H_1 and X_1 on wye-delta transformers would not be followed.

  (C) A positive-sequence voltage drop from Bushing H_1 to the neutral would lead the positive-sequence voltage drop from Bushing X_1 to the neutral by more than 30°.

  (D) The delta winding must be located on the high-voltage side of the transformer.

GO ON TO THE NEXT PAGE

503. An industrial plant is served by a 12,470-V Δ/480-V grounded-wye transformer. The high-voltage side of the transformer is served from a 12,470-V wye-connected distribution line that has a grounded neutral. The best protection of the transformer against lightning strikes on the 12,470-V line will result from surge arrester connections of:

(A) phase to ground on the 12,470-V side of the transformer

(B) phase to phase on the 12,470-V side of the transformer

(C) phase to ground on the 480-V side of the transformer

(D) phase to neutral to ground on the 480-V side of the transformer

504. Answer this question in accordance with the 2002 edition of the *National Electrical Code®* *(NEC®).*[1]

The minimum size THWN copper conductors rated at 75°C installed in conduit required to serve a continuous duty 230-V 10-hp single-phase induction motor and non-continuous 1-kW resistance heater on a circuit operating at 240-V single phase are:

(A) 4/0 AWG
(B) 2/0 AWG
(C) 4 AWG
(D) 6 AWG

---

[1] *National Electrical Code®* and *NEC®* are registered trademarks of the National Fire Protection Association, Inc., Quincy, MA 02269.

POWER AFTERNOON SAMPLE QUESTIONS

505. A squirrel-cage induction motor is rated 40 hp, 460 V, 3 phase, 60 Hz, 52 A, 0.87 lagging power factor at full load, **National Electrical Code**® Design E (**National Electric Code**® 2002). The maximum locked-rotor current (amperes) for selection of the motor disconnect is most nearly:

(A) 412
(B) 356
(C) 300
(D) 240

506. You have sized branch-circuit conductors supplying a 3-phase squirrel-cage induction motor in accordance with the 2002 edition of the **National Electrical Code**®. Now you must size the fuses to protect those conductors against short circuit or ground faults. What is the correct approach?

(A) The maximum fuse rating is the ampacity of the conductors, except the next larger size fuses can be used if the conductor ampacity does not match a standard fuse size.

(B) The maximum fuse rating is 125% of the full-load current of the motor.

(C) The maximum fuse rating is 100% of the locked-rotor current of the motor.

(D) None of the above.

POWER AFTERNOON SAMPLE QUESTIONS

507. A 3-phase transmission line is rated 65 kV and 24 MVA. The line impedance is 50 Ω/phase. Assuming the line-rated values are also the base values, the per-unit impedance is most nearly:

(A) 0.167
(B) 0.284
(C) 3.52
(D) 18.46

508. A 3-phase, 4-wire, neutral-grounded, wye-connected utility line has a phase-to-phase voltage of 13.2 kV. A complex load of $(200 + j100)$ kVA is connected between Phase A and neutral. An identical load is connected between Phase B and neutral. The neutral current (amperes) is most nearly:

(A) 0
(B) 9.8
(C) 16.9
(D) 29.3

POWER AFTERNOON SAMPLE QUESTIONS

509. A 3-phase, 3-wire, ungrounded, 13.2-kV (phase-to-phase) wye-connected source is connected to a balanced delta load that is grounded on Corner A. The voltage measured between Corner B and ground is most nearly:

(A) half the phase-to-phase voltage
(B) 7.62 kV
(C) 13.2 kV
(D) cannot be determined

510. The only load on a 3-phase, 4-wire system is placed between Phase B and Phase C. The phase-to-phase voltage is 13.2 kV. The load is 500 kVA at 0.85 lagging power factor. The magnitude of the line current in Phase C (amperes) is most nearly:

(A) 65.6
(B) 55.8
(C) 37.9
(D) 32.2

GO ON TO THE NEXT PAGE

511. The figure below represents a balanced 3-phase distribution system, sequence ABC, 60 Hz.

$V_{AB} = 12.5 \angle 0°$ kV (load voltage), and the impedance of the line is $(5 + j10)$ Ω/phase.

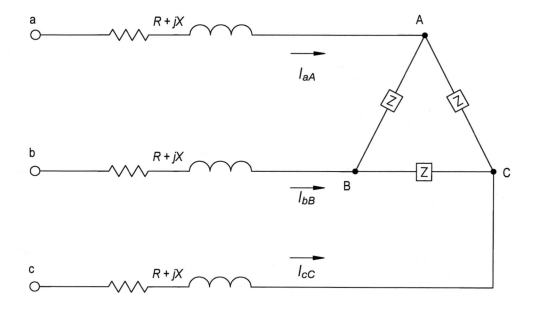

If $I_{aA} = 70 \angle -20°$, then the magnitude $|V_{ab}|$ in kV is most nearly:

(A) 15.0
(B) 13.8
(C) 13.0
(D) 11.9

GO ON TO THE NEXT PAGE

512. When using the method of symmetrical components, which of the following statements is most nearly correct?

(A) The three zero-sequence current phasors are equal in magnitude and displaced by $120°$.

(B) The three positive-sequence and the three negative-sequence current phasors are all equal in magnitude, but have opposite phase rotations.

(C) In a wye-connected synchronous generator, the voltage drop through the neutral grounding resistor, R, is equal to $\sqrt{3} \, I_0 R$, where I_0 is the zero-sequence current.

(D) Positive-sequence and negative-sequence currents cannot pass through the neutral grounding resistor in a wye-connected generator.

POWER AFTERNOON SAMPLE QUESTIONS

513. Consider a 208/120 V, 3-phase, wye-connected, 4-wire transformer. Each 120-V phase feeds a separate and identical dc power supply comprising a simple full-wave bridge rectifier, followed by a large filter capacitor. There are a total of three dc power supplies, one connected to each phase. The following circuit diagram shows one of the three power supplies:

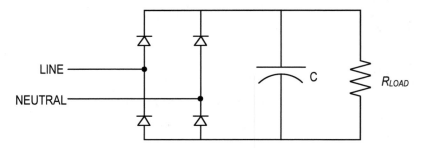

The outputs of the three power supplies are loaded with identical resistive loads. There are no other loads on the transformer.

Which of the following oscilloscope traces of the neutral current is most likely to be observed?

(A)
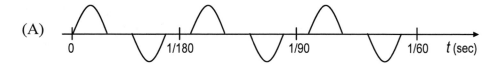

(B)

(C)
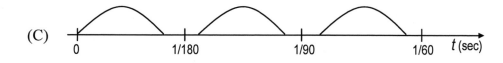

(D) The trace is essentially flat.

514. The following inverter schematic is used for a variable-speed motor drive. Assume ideal transistors. Each transistor operates as an ideal switch.

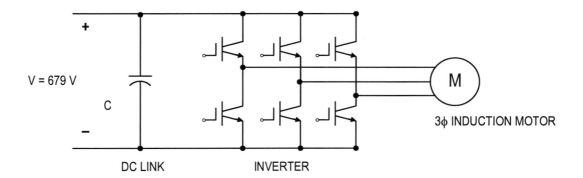

The peak phase-to-phase voltage (V) to the motor is most nearly:

(A) 336
(B) 475
(C) 480
(D) 679

515. A single-phase, 60-Hz, fully controlled thyristor bridge operates with a purely resistive load. In the figure shown below, i_a is the output current, v_a is the output voltage, and the input current is designated as i_s.

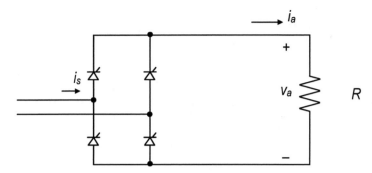

The output waveform is illustrated here:

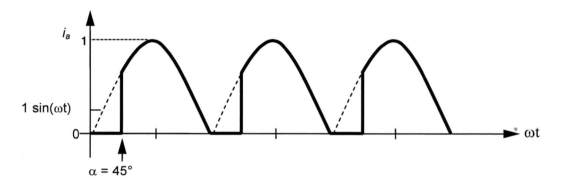

The average value of the output current i_a (amperes) is most nearly:

(A) 0
(B) 0.543
(C) 0.637
(D) 0.707

516. Consider the three-legged magnetic circuit in the figure, fabricated with a homogeneous iron core and a uniform cross-sectional area. With the coil energized, the magnetic flux in Leg A is 3.0×10^{-3} webers (Wb). Assuming linear magnetic properties, the approximate flux (Wb) in Leg C is most nearly:

(A) 0.5×10^{-3}
(B) 0.75×10^{-3}
(C) 1.0×10^{-3}
(D) 2.0×10^{-3}

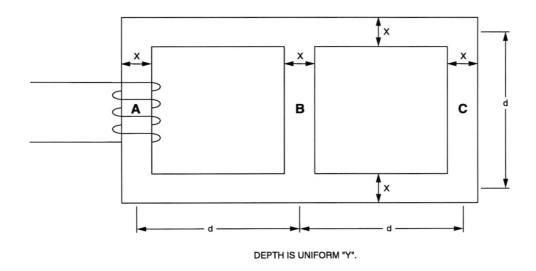

DEPTH IS UNIFORM "Y".

517. A 3-phase distribution line conductor configuration is as follows:

•←3 ft→•←4 ft→•

The 12-kV distribution line conductor geometric mean distance (feet) for the crossarm framing shown is most nearly:

(A) 3.5
(B) 4.0
(C) 4.4
(D) 5.0

POWER AFTERNOON SAMPLE QUESTIONS

518. A 3-phase, 60-Hz transmission system is shown in the figure below. The transmission line current is 75.93 A.

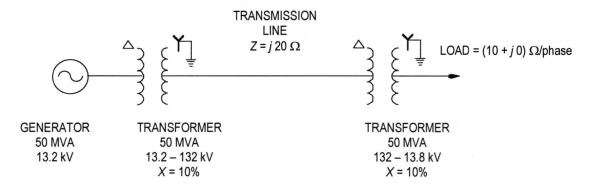

The load current (amperes) in the generator is most nearly:

(A) 760
(B) 1,320
(C) 2,090
(D) 2,190

519. An induction motor is rated at 1,000 hp, 1,165 rpm, 2.4 kV, 201 A, 3-phase, 60 Hz, 0.94 lagging power factor at full load.

If the starting torque of the motor is 150% of full-load torque, then the starting torque (ft-lb) is most nearly:

(A) 4,510
(B) 5,250
(C) 6,020
(D) 6,760

POWER AFTERNOON SAMPLE QUESTIONS

520. A dc shunt motor has a nameplate rating of 150 hp, 600 V, 200 A, 1,750 rpm.

Assuming a no-load speed of 1,790 rpm, the speed regulation is most nearly:

(A) 2.2%
(B) 2.3%
(C) 2.5%
(D) 3.0%

521. A synchronous generator is 3-phase, wye-connected, and rated 1,131 kVA, 2.4 kV, 0.90 lagging power factor.

The output current (amperes) from the generator at rated full load is most nearly:

(A) 245
(B) 272
(C) 424
(D) 471

522. A power plant uses a turbine-driven synchronous generator rated 3-phase, 150 MVA, 13.8 kV, 0.85 lagging power factor.

At rated conditions the real power output (MW) is most nearly:

(A) 60
(B) 75
(C) 128
(D) 150

GO ON TO THE NEXT PAGE

523. A 3-phase induction motor draws 28.0 A at no-load and 92.0 A at full-load. The power factor of the motor at half-load is most nearly:

(A) 0.61
(B) 0.64
(C) 0.71
(D) 0.84

524. A large 3-phase induction motor is to be specified. Which of the following design features would be most helpful in reducing peak starting-current requirements?

(A) Use of a high-efficiency, Design E motor

(B) Starting the motor unloaded and then connecting the driven device with a clutch after the motor has accelerated

(C) Use of a motor with a higher service factor

(D) Use of a wye-start, delta-run motor

POWER AFTERNOON SAMPLE QUESTIONS

525. Two single-phase transformers with identical voltage ratings are proposed to be connected in parallel to serve a load. The following rating and impedance data are provided:

Transformer 1: 1,000 kVA, $Z = 4.5\%$
Transformer 2: 2,000 kVA, $Z = 6.0\%$

The maximum total load (kVA) that can be served by the bank without overloading either transformer is most nearly:

(A) 2,333
(B) 2,500
(C) 2,667
(D) 3,000

526. Two 3-phase transformers are proposed to be connected in parallel to serve a large load. Transformer 1 is wye-delta, and Transformer 2 is wye-wye. Which of the following statements is most correct?

(A) The transformers should not be paralleled.

(B) The transformers may be paralleled if the neutral of Transformer 2 secondary is unconnected and the turns ratio of Transformer 2 is 1.732 times the turns ratio of Transformer 1.

(C) The transformers may be paralleled if the neutral of Transformer 2 secondary is unconnected and the turns ratio of Transformer 1 is 1.732 times the turns ratio of Transformer 2.

(D) The secondary neutrals of Transformer 1 and 2 may or may not be connected, but the transformer turns ratios, impedances, and kVA ratings must be matched to achieve maximum output.

527. A transformer is rated 3-phase, 25 MVA, 66-12.47 kV, delta-wye connected, 8% impedance. On a 100-MVA base, the per-unit impedance is most nearly:

(A) 0.08
(B) 0.16
(C) 0.24
(D) 0.32

528. Which of the following conditions would most likely result in a saturated magnetic flux within the indicated equipment?

(A) A 50-Hz power transformer operated at 60 Hz and at rated MVA

(B) A 3-phase induction motor driven by an inverter at the rated voltage of the motor and half its rated frequency

(C) A power transformer, operating at rated MVA, connected delta to a long transmission line that is exposed to a heavy geomagnetic solar storm

(D) A current transformer with its primary carrying rated current and its secondary winding shorted

529. Answer this question in accordance with the 2002 edition of the *National Electrical Code®*. Assume the voltage at the main distribution panel is 480 V, 3-phase. Feeder F1 consists of three 500-kcmil THWN copper phase conductors and a neutral in a steel conduit that runs a distance of 250 feet. Panel A serves a constant, balanced load of 400 A at 0.80 lagging pf. The voltage (V) at Panel A is most nearly:

(A) 475
(B) 470
(C) 465
(D) 460

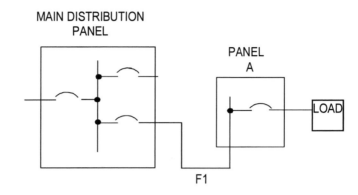

530. To correct power factor, a capacitor is proposed to be connected to Point A in the 3-phase induction motor circuit shown in the figure below.

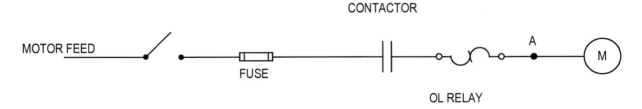

Which of the following statements is most likely correct if the capacitor is added?

(A) The full-load current at Motor M will be significantly reduced.

(B) The motor overload relay trip settings should be reduced.

(C) The 2002 edition of the *National Electrical Code®* requires that the capacitor be connected through a separate disconnect switch.

(D) The motor feed conductors may be sized for a reduced ampacity.

POWER AFTERNOON SAMPLE QUESTIONS

531. A 3-phase capacitor rated at 240 V and 110 kvar has been proposed for correcting the power factor of a 3-phase induction motor operating at 208 V. The kvar that will be provided by the capacitor is most nearly:

(A) 82.6
(B) 95.3
(C) 110
(D) 127

532. Consider the 60-kV transmission system below. Transmission line impedances are:

$$Z_1 = 16.75\angle71° \, \Omega$$
$$Z_2 = 13.4\angle71° \, \Omega$$

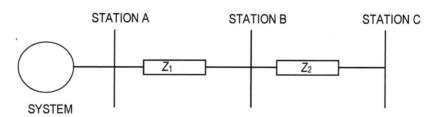

With a system impedance of $13.25\angle81° \, \Omega$, the 3-phase fault current (amperes) at Station C is most nearly:

(A) 2,590
(B) 1,495
(C) 1,285
(D) 800

533. The maximum resistance to ground (Ω) of a single grounding electrode without the requirement to add an additional grounding electrode that is found in the 2002 edition of the *National Electrical Code®* is:

(A) 5
(B) 20
(C) 25
(D) 50

534. The circuit represented in the figure below is a ground-fault protection scheme.

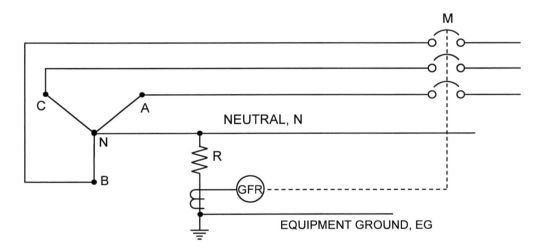

The purpose of R is to:

(A) limit the ground current and prevent damage

(B) maintain a potential difference between N and EG

(C) decrease the interrupting rating required of M

(D) A, B, and C are correct

535. An electric generation facility uses a turbine-driven synchronous generator rated 3-phase, 150 MVA, 13.8 kV.

Per-unit (pu) reactances are $X''_d = 0.15$; $X'_d = 0.25$; $X_d = X_q = 1.20$

Assume terminal voltage $(E_t) = 1.0$ pu

For a simple transient study at rated MVA, rated voltage, and unity power factor, the internal voltage (pu) and reactance (pu) for the generator should be, respectively:

(A) 1.01 and 0.15
(B) 1.01 and 0.25
(C) 1.03 and 0.25
(D) 1.10 and 0.25

536. Consider a long 3-phase feeder with the following phase conductor properties:

Reactance, $X_L = 0.050 \ \Omega$
ac resistance, $R = 0.050 \ \Omega$

The feeder is supplying a balanced load. For a fixed current of 50 A, which of the following statements concerning the phase-to-neutral voltage drop at the receiving end of the feeder is most correct?

(A) The voltage drop is relatively independent of the load power factor.

(B) The voltage drop will be largest for a unity load power factor.

(C) The voltage drop will be largest for a zero load power factor.

(D) The voltage drop will be larger for a lagging load power factor of 0.707 than for either a unity load power factor or a zero load power factor.

GO ON TO THE NEXT PAGE

GO ON TO THE NEXT PAGE

537. The figure below represents a 115-kV transmission system serving a 115-12.47-kV, 3-phase distribution transformer. The 12-kV OCBs serve loads only, no sources.

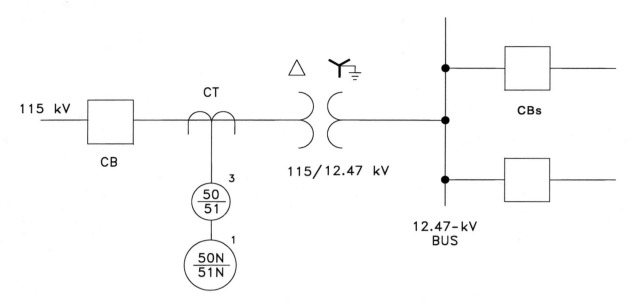

A 3-phase short circuit on the 12.47-kV bus results in a short-circuit current of 1,500 A on the 115-kV system. Relay 50/51, with characteristics as shown in the following graph, has a minimum pick-up of 420 A. For relay operation in 0.9 second, the time dial setting is:

(A) 5
(B) 3
(C) 2
(D) 1/2

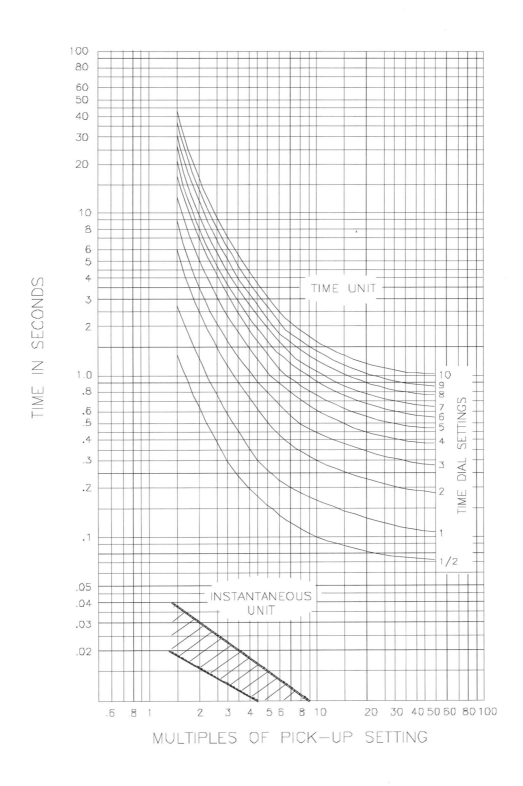

GO ON TO THE NEXT PAGE

538. Assuming that only the following standard fuse sizes are available, what is the maximum size (ampere) overcurrent device normally allowed to protect AWG No. 10 THWN copper conductors per the 1999 edition of the *National Electrical Code®*?

(A) 30
(B) 35
(C) 40
(D) 35 unless connectors are rated at 60°C, in which case 30.

539. An overcurrent relay is used to protect a circuit with a maximum 3-phase fault current of 8,000 A primary. A 400:5 current transformer ratio and a 5-A relay tap setting are selected. If the total secondary burden on the current transformer for the maximum fault condition is 1.10 Ω, the excitation voltage (V) that must be developed by the current transformer is most nearly:

(A) 100
(B) 110
(C) 115
(D) 120

540. If the setpoint of relay CV is 5 V, its contacts will change state when the value (kV) of V_{NG} is most nearly:

(A) 0.6
(B) 0.5
(C) 0.4
(D) 0.2

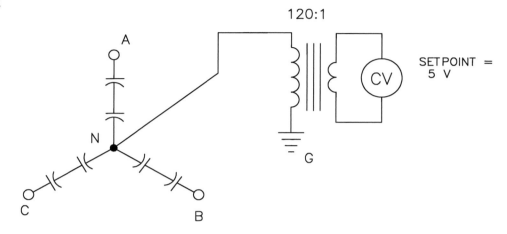

ELECTRICAL AND COMPUTER ENGINEERING
BREADTH MORNING SAMPLE SOLUTIONS

CORRECT ANSWERS TO THE ELECTRICAL AND COMPUTER MORNING SAMPLE QUESTIONS

Detailed solutions for each question begin on the next page.

101	A	**121**	B
102	C	**122**	D
103	D	**123**	B
104	A	**124**	A
105	D	**125**	A
106	B	**126**	B
107	A	**127**	B
108	B	**128**	C
109	D	**129**	D
110	D	**130**	A
111	D	**131**	C
112	C	**132**	D
113	B	**133**	B
114	B	**134**	C
115	B	**135**	B
116	C	**136**	C
117	A	**137**	A
118	D	**138**	C
119	A	**139**	B
120	D	**140**	D

101. With no consideration for the cost of capital, the payout rate ($/mo) is simply the Benefits minus Costs. Then the payout time is the One-time Cost divided by the Payout Rate.

Vendor	One-time Cost ($)	Benefits ($/mo)	Total Costs ($/mo)	Net Benefits ($/mo)	Payout Time (mo)
A	24,000	4,000	1,600	2,400	10.0
B	27,000	4,200	1,700	2,500	10.8
C	30,000	4,500	1,700	2,800	10.7
D	32,000	4,800	1,900	2,900	11.0

Vendor A provides the lowest payout time and would therefore be chosen based on this criterion.

THE CORRECT ANSWER IS: (A)

102.

$$PW = 50,000\left[1+\left(\frac{P}{F}\right)_2^{10\%}\right]+20,000\left(\frac{P}{A}\right)_4^{10\%}-10,000\left(\frac{P}{A}\right)_2^{10\%}$$

$$= 50,000\left[1+(1+0.1)^{-2}\right]+20,000\left(\frac{(1+0.1)^4-1}{(0.1)(1+0.1)^4}\right)-10,000\left(\frac{(1+0.1)^2-1}{(0.1)(1+0.1)^2}\right)$$

$$= 50,000\left[1+0.82645\right]+20,000\left(\frac{1.4641-1}{(0.1)(1.4641)}\right)-10,000\left(\frac{1.21-1}{(0.1)(1.21)}\right)$$

$$= 50,000(1.82645)+20,000(3.1699)-10,000(1.73554)$$

$$= 91,323+63,398-17,355$$

$$= 137,365$$

THE CORRECT ANSWER IS: (C)

103.

$$P_1 = (1-R_1) = (1-0.92) = 0.08$$
$$P_2 = (1-R_2) = (1-0.85) = 0.15$$

$$R_1 \| R_2 = 1-(P_1 \times P_2)$$
$$= 1-(0.08)(0.15)$$
$$= 1-0.012$$
$$= 0.988$$

THE CORRECT ANSWER IS: (D)

ELECTRICAL AND COMPUTER MORNING SAMPLE SOLUTIONS

104. The grounded neutral conductor should be electrically bonded to the grounding electrode conductor and the equipment grounding conductor at the service entrance or source of a separately derived system. (B) is not correct since this connection can cause inappropriate activation of GFI devices. (C) is not correct since this connection can cause objectionable neutral currents to flow through the equipment grounding system.

THE CORRECT ANSWER IS: (A)

105. Of the choices, the line-to-neutral shock is least protected. (A) is not the best choice since a short circuit is protected by the circuit breaker. (B) is not the best choice since an overload is protected by the circuit breaker. (C) is not the best choice since a high-resistance phase-to-ground fault is protected by the GFCI.

THE CORRECT ANSWER IS: (D)

106. Use Norton's Theorem. The conductance will be:
$$G = \frac{(3.0 - [5.0][0.2])}{5.0} = \frac{2.0}{5.0} = 0.4 \text{ S}$$
The parallel current source will have a value of 3.0 A.

THE CORRECT ANSWER IS: (B)

107. Since the frequencies of the two components of the current are different, the total power may be calculated as:
$$P = \left(\frac{5}{\sqrt{2}}\right)^2 (3) + \left(\frac{2}{\sqrt{2}}\right)^2 (3)$$
$$= \left(\frac{25}{2}\right)(3) + \left(\frac{4}{2}\right)(3)$$
$$= 43.5 \text{ W} \cong 44 \text{ W}$$

THE CORRECT ANSWER IS: (A)

108. First calculate the magnitude of the voltage:

$$V = \frac{50\angle 45°}{\sqrt{2}} + \frac{28.28\angle 0°}{\sqrt{2}}$$

$$= \left(\frac{50}{\sqrt{2}}\right)(0.707 + j0.707) + 20$$

$$= 25 + j25 + 20$$

$$= 45 + j25$$

$$= 51.5\angle 29° \text{ V}$$

The rms value of the voltage is 51.5 V. The power is computed by:

$$P = \frac{V^2}{R} = \frac{(51.5)^2}{5} = 530 \text{ W}$$

THE CORRECT ANSWER IS: (B)

109. At $t = 0$,

$$v(0) = 6\left(\frac{2k}{3k}\right) = 4 \text{ V}$$

Given $R = 2 \text{ k}\Omega$, $C = 3 \times 10^{-6} \text{ F}$, thus $\tau = RC = 2,000 \times 3 \times 10^{-6} = 6 \text{ ms}$
$v(0) = A + B = 4$
$v(\infty) = A = 0$
Therefore, $B = 4$

THE CORRECT ANSWER IS: (D)

110. The waveform has a zero average value, so the dc component is zero.

$v(t) = -v(-t)$, so the function is odd, not even, and contains sines.

THE CORRECT ANSWER IS: (D)

111. $I_1(s) = I(s) \dfrac{R}{R + R_1 + \dfrac{1}{sC} + sL}$

$V(s) = I_1(s) \cdot (sL) = \dfrac{R(sL)(sC)I(s)}{s^2 LC + (R + R_1)sC + 1} = \dfrac{s^2 RI(s)}{s^2 + \dfrac{(R + R_1)}{L}s + \dfrac{1}{LC}}$

$\dfrac{V(s)}{I(s)} = \dfrac{s^2 R}{s^2 + \dfrac{(R + R_1)}{L}s + \dfrac{1}{LC}}$

THE CORRECT ANSWER IS: (D)

112. Using Thevenin's theorem:

$V_{TH} = V_{OC} = 100\angle 0° \text{ V}$

$Z_{TH} = \dfrac{V_{OC}}{I_{SC}} = \dfrac{100\angle 0°}{5\angle 45°}$

$\quad = 20\angle -45° = (14.14 - j14.14)\Omega$

The best answer is: $100\angle 0°$ V and $(14 - j14)$ Ω.

THE CORRECT ANSWER IS: (C)

113. $I_A = 5/2 = 2.5 \text{ A}$
$G_B = 1/R_B = 0.25 \text{ S}$

THE CORRECT ANSWER IS: (B)

114. $I(s) = \dfrac{50/s}{10 + 5 \times 10^{-3}s + \dfrac{1}{s \times 200 \times 10^{-6}}}$

$= \dfrac{50/(5 \times 10^{-3})}{s^2 + \dfrac{10s}{5 \times 10^{-3}} + \dfrac{1}{(200 \times 10^{-6}) \times (5 \times 10^{-3})}}$

$= \dfrac{10{,}000}{s^2 + 2{,}000s + 10^6}$

$= \dfrac{10{,}000}{(s + 1{,}000)(s + 1{,}000)}$

$= \dfrac{10{,}000}{(s + 1{,}000)^2}$

THE CORRECT ANSWER IS: (B)

115. For maximum power transfer, the impedance of the load after transformation should be the same as the impedance of the source. For an ideal transformer, $\dfrac{Z_1}{Z_2} = \left(\dfrac{N_1}{N_2}\right)^2$

$\dfrac{N_1}{N_2} = \sqrt{\dfrac{Z_1}{Z_2}} = \sqrt{\dfrac{90}{15}} = \sqrt{6} = 2.449$

THE CORRECT ANSWER IS: (B)

116. $H = \dfrac{I}{2\pi R} = \dfrac{50}{0.2\pi} = 79.6 \text{ A/m}$

THE CORRECT ANSWER IS: (C)

117. Two *AND* gates into an *OR* gate yields
$AB + CB = B(A+C)$

THE CORRECT ANSWER IS: (A)

118. Simple conversion from POS to SOP, can be accomplished in any number of ways. Here is one of them:

$$(\overline{A} + B)(\overline{B} + \overline{C})(A + C)$$
$$= (\overline{A}\,\overline{B} + \overline{A}\,\overline{C} + B\overline{B} + B\overline{C})(A + C)$$
$$= (\overline{A}\,\overline{B}A + \overline{A}\,\overline{B}C + \overline{A}\,\overline{C}A + \overline{A}\,\overline{C}C + B\overline{C}A + B\overline{C}C)$$
$$= \overline{A}\,\overline{B}C + AB\overline{C}$$

THE CORRECT ANSWER IS: (D)

119. $R = V/I$ at the peak. $R = 169$ V/5 A = 33.8 Ω.

THE CORRECT ANSWER IS: (A)

120. The circuit transfer function (with $s = j\omega$), has a magnitude of

$$|A(j\omega)| = \left| \frac{\dfrac{1}{j\omega C}}{R} \right| = \left| \frac{1}{RCj\omega} \right| = \frac{1}{RC\omega}$$

This is a straight line in Bode plot terms, with the magnitude decreasing based on the frequency ω. [Note $s = j\omega = j(2\pi f)$]. The 0 dB point (unity gain) occurs when $\omega = \omega_c = 1/RC$. These characteristics are present in (D).

THE CORRECT ANSWER IS: (D)

121. If $V_C = 10$ V, then the drop across the 5-kΩ resistor is 10 V, giving a collector current $I_C = 2$ mA. Since $\beta = 100$, then $I_E \approx 2$ mA. Then $V_E = I_E R_E = (2$ mA$)(1,000$ $\Omega) = 2$ V.

THE CORRECT ANSWER IS: (B)

122. Since the capacitor charges to 220×1.4 V, and holds that charge while the supply goes to its negative maximum -220×1.4, the PIV must be $440 \times 1.4 = 616$ V.

THE CORRECT ANSWER IS: (D)

123. The power input to the speaker is $V^2/R = (30)^2/8 = 900/8 = 112.5$ W. The speaker output is $(112.5$ W$)(0.15) = 17$ W.

THE CORRECT ANSWER IS: (B)

124. $I_E = \dfrac{10 - 0.7}{20k} = 0.465 \text{ mA}$

$\alpha = \dfrac{\beta}{\beta + 1} = \dfrac{100}{101} \cong 0.99$

$I_C = \alpha I_E = (0.99)(0.465) = 0.4604 \cong 0.46 \text{ mA}$

$v_C = -10 + I_C R_C = -10 + (0.46 \cdot 10^{-3})(10 \text{ k})$

$\quad\quad = -10 + 4.6 = -5.4 \text{ V}$

A positive swing of 5.4 V will result in saturation of the BJT.
A negative swing of 4.6 V will result in clipping of the output signal.
The maximum swing is therefore limited to the smaller value of 4.6 V.

THE CORRECT ANSWER IS: (A)

125. Conductivity is given by $G = q(\mu_p + \mu_N)n_i$ where q is the charge, μ is the mobility, and n_i is the intrinsic charge. The conductivity of a heavily doped n-type silicon sample is proportional to the electron mobility times the electron concentration. However, the conductivity is weakly dependent on the electron mobility when compared to the electron concentration..

THE CORRECT ANSWER IS: (A)

126. Output waveform:

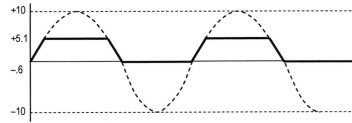

$v_o = \dfrac{5.1}{2} = 2.55 \text{ V}$ if a rectangular approximation is used. The actual average voltage will be slightly less.

THE CORRECT ANSWER IS: (B)

127. Open-loop transfer function = $50/[40s(0.5s + 1)]$
Closed-loop transfer function = $2.5/(s^2 + 2s + 2.5)$
$\omega_n = 1.58$
$2\xi\omega_n = 2$
$\xi = 1/\omega_n = 0.633$

THE CORRECT ANSWER IS: (B)

128. Each zero contributes +20 dB/dec.
Each pole contributes –20 dB/dec.
The pole-zero excess is 4 – 2 = 2, so the high-frequency slope is – 40 dB/dec.

THE CORRECT ANSWER IS: (C)

129. A digital system is stable if the poles are inside the unit circle and the region of convergence includes the unit circle. The system becomes less stable as the poles are moved closer to the unit circle.

THE CORRECT ANSWER IS: (D)

130. $\dfrac{C}{R} = \dfrac{G}{1+G} = \dfrac{N(s)}{D(s)}$

$\begin{aligned} D(s) &= (s+1)(s+2) + K \\ &= s^2 + 3s + (2+K) \end{aligned}$

Therefore the system is stable only for $K > -2$

THE CORRECT ANSWER IS: (A)

131. An amplitude signal contains both carrier and upper and lower sidebands.

THE CORRECT ANSWER IS: (C)

132. The phase margin is the number of degrees above –180° at the gain crossover point. For the system shown in the figures, the phase margin equals 90° at the gain crossover point.

THE CORRECT ANSWER IS: (D)

133. $V_{sec} = I_{line}\left(\dfrac{0.78 + j0.052}{1,000 \text{ ft}}\right)(200 \text{ ft}) + V_{load} + I_{line}\left(\dfrac{0.78 + j0.052}{1,000 \text{ ft}}\right)(200 \text{ ft})$

$\quad = 240\angle 0° + (2)(30\angle 0°)\left(\dfrac{1}{5}\right)(0.78 + j0.052)$

$\quad = 249.4\angle 0.143° \text{ V}$

$\left|V_{sec}\right| = 249.4 \text{ V}$

THE CORRECT ANSWER IS: (B)

134. $S_{motor} = \sqrt{3}V_lI_l\angle\cos^{-1}(pf) = \sqrt{3}(480)(34)\angle\cos^{-1}(0.75) = 28.3\angle 41.4°\,\text{kVA}$

$\begin{aligned} Q_{new} &= P[\tan(\cos^{-1}\theta)] \\ &= 21.7[\tan(\cos^{-1}0.9)] = 10.3 \end{aligned}$

$Q_{new} = Q_{old} + Q_{add}$

$\begin{aligned} Q_{add} &= Q_{new} - Q_{old} = 10.3 - 18.7 \\ &= -8.4\,\text{kvar} \end{aligned}$

THE CORRECT ANSWER IS: (C)

135. $P_{load} = 8,000\,\text{kW}$

$\text{P.F.} = 0.80$

$S_{load} = \dfrac{8,000\,\text{kW}}{0.80} = 10,000\,\text{kVA}$

$Q_{load} = \sqrt{10,000^2 - 8,000^2} = 6,000\,\text{kvar}$

$S_{new} = \dfrac{8,000\,\text{kW}}{0.95} = 8421\,\text{kVA}$

$Q_{new} = \sqrt{8,421^2 - 8,000^2} = 2,630\,\text{kvar}$

$Q_{cap} = Q_{load} - Q_{new} = 6,000 - 2,630 = 3,370\,\text{kvar}$

THE CORRECT ANSWER IS: (B)

136. $\left|I_F\right| = \left|\dfrac{V}{Z}\right| = \left|\dfrac{V}{jX_S}\right| = 3,600\,\text{A}$

$X_S = \dfrac{7,200\,\text{V}}{3,600\,\text{A}} = 2\,\Omega$

When the fault resistance increases to 1 Ω, the new fault current will be:

$\left|I_F\right| = \left|\dfrac{V}{R_F + jX_S}\right| = \left|\dfrac{7,200}{1+j2}\right| = \dfrac{7,200}{\sqrt{1^2 + 2^2}} = 3,220\,\text{A}$

THE CORRECT ANSWER IS: (C)

137. The equipment grounding conductor provides a low-impedance path to ground to ensure proper ground-fault clearing. It is connected to the grounded neutral and the grounding electrode conductor only at the service entrance or the source of a separately derived system and thus does not provide a path for triplen harmonics.

THE CORRECT ANSWER IS: (A)

138. The line voltages on the high and low sides are 115 kV and 24 kV respectively. Because of the connection, the phase voltage on the high side is 115 kV, and the phase voltage on the low side is 24 kV/1.732 = 13.86 kV.

The turns ratio is 115/13.86 = 8.3

THE CORRECT ANSWER IS: (C)

139. Synchronous speed is given by 120 f/P, where f is the frequency (Hz) and P is the total number of poles. An induction motor will slip slightly when operating under load. Therefore it will run at a speed slightly less than synchronous. The synchronous speed just above 1,600 rpm that corresponds to 60 Hz and an even number of poles is 1,800 rpm. This implies that the number of poles is four.

THE CORRECT ANSWER IS: (B)

140.
$$Z_{high} = \left(\frac{N_{high}}{N_{low}}\right)^2 Z_{low} = \left(\frac{2}{1}\right)^2 (4\angle 30°) = 16\angle 30° \; \Omega$$

THE CORRECT ANSWER IS: (D)

DEPTH AFTERNOON SAMPLE SOLUTIONS

ELECTRICAL AND COMPUTER ENGINEERING

COMPUTERS
AFTERNOON SAMPLE SOLUTIONS

CORRECT ANSWERS TO THE COMPUTERS AFTERNOON SAMPLE QUESTIONS

Detailed solutions for each question begin on the next page.

501	B	**521**	B
502	A	**522**	D
503	A	**523**	C
504	A	**524**	A
505	C	**525**	D
506	D	**526**	C
507	A	**527**	B
508	A	**528**	B
509	A	**529**	B
510	A	**530**	B
511	D	**531**	B
512	C	**532**	C
513	D	**533**	B
514	A	**534**	D
515	B	**535**	B
516	D	**536**	B
517	C	**537**	C
518	C	**538**	D
519	C	**539**	C
520	B	**540**	C

COMPUTERS AFTERNOON SAMPLE SOLUTIONS

501. In the context of the EIA RS-232-C specification, the maximum length for a DTE cable is 15 meters.

THE CORRECT ANSWER IS: (B)

502. The start bit is indicated by a negative-to-positive voltage transition at T0. The data stream indicates 8 data bits (T1-T9) and a trailing parity bit (T10), giving the total data frame (T1-T10) even parity. Other selections fail to consider the parity bit or fail to properly calculate the parity bit.

THE CORRECT ANSWER IS: (A)

503. The bit format for the floating point number is as follows: $S=(0)_2$ $E=(1000000)_2$ $M=(00000000)_2$. The sign bit S is 0, indicating a positive number. The exponent field has value 64, when converted from excess-64 yields 0 as the true exponent value. The mantissa is zero, so the data value is $0 \times 2^0 = 0.0$, or value $= -1^S (M * 2^{E-64}) = (+1)(0.0 * 2^{64-64}) = 0.0$

THE CORRECT ANSWER IS: (A)

504. Shifting left by 3 bits is equivalent to multiplying by 8.

THE CORRECT ANSWER IS: (A)

505. The step size is 10 V/16 = 0.625 V. Since the encoding of the A/D converter is offset by –5 V, the true voltage magnitude being encoded is 2 V – (–5 V) = 7 V. 7 V/0.625 V = 11.2, which is $(1011)_2$.

THE CORRECT ANSWER IS: (C)

506. The required data at the latch is $(01xxxx10)_2$. The latch is addressed at location $(111x\ xxxx\ xxxx\ xxxx)_2$.

THE CORRECT ANSWER IS: (D)

507. U4 is driven by U1 and U2. Either U1 or U2 sets an upper limit on this propagation delay.

THE CORRECT ANSWER IS: (A)

508. The output is GT (greater than) if the previous stage indicated GT OR the previous stage is NOT LT (less than) AND A is greater than B (A=1 and B=0).

THE CORRECT ANSWER IS: (A)

509. The truth table is as follows:

A = 0, B = 0 → Z = 1
A = 0, B = 1 → Z = 0
A = 1, B = 0 → Z = 0
A = 1, B = 1 → Z = 0

This corresponds to a NOR gate function.

THE CORRECT ANSWER IS: (A)

510. When C = 0, the lower chain of NMOS devices is disabled, and the upper chain is enabled. When A and B are 1, the output is 1 as the NMOS upper devices are enabled through the control PMOS device. When A = 0, B passes through the lower transmission gate to the output, and when B = 0, A passed through the upper transmission gate to the output. Thus, if either A or B is a logic 1 and the other input is a logic 0, the output is a logic 1. The truth table is as follows:

A = 0, B = 0, Y = 0
A = 0, B = 1, Y = 1
A = 1, B = 0, Y = 1
A = 1, B = 1, Y = 1

This corresponds to an OR function.

THE CORRECT ANSWER IS: (A)

511. The frame contains 11 bits total (1 start bit, 9 data bits, and 1 stop bit). 11/9,600 = 1.145 ms or 1,146 microseconds.

THE CORRECT ANSWER IS: (D)

512. Since X is asynchronous to the other inputs, any output state is possible including 11. The asynchronous input will occasionally violate the setup and hold times of the two flip-flops. In this case, neither may respond, both may respond, or just one may respond. Possible transitions are 01 → 01 (X = 0); 01 → 10 (X = 1); 01 → 00, 01 → 11 (X = 1, violation of setup/hold times on one FF with the other responding).

THE CORRECT ANSWER IS: (C)

COMPUTERS AFTERNOON SAMPLE SOLUTIONS

513. The output M will not change state until the *rising* edge of the clock. This eliminates responses A and C. The *next* rising edge will ripple the Q output to the state machine, which will feed back a logic 1 to the asynchronous clear.

THE CORRECT ANSWER IS: (D)

514. M must immediately latch a 1 on the rising edge of the 40-MHz pulse. This eliminates responses B and D. N must be latched high on the next rising edge of the 10-MHz clock and remain high until the next rising edge of the 10-MHz clock, since M is cleared asynchronously immediately once N goes high.

THE CORRECT ANSWER IS: (A)

515. The system should transition by nickel increments and dispense candy when $0.25 has been dispensed. Responses B and D are the only state diagrams satisfying this requirement. Response D fails when N, N, N, N, D sequence is used. Also when N, N, N, D is used.

THE CORRECT ANSWER IS: (B)

516. The state transition table for a JK flip-flop is as follows:

J	K	OUTPUT
0	0	hold
0	1	0
1	0	1
1	1	toggle

The J/K inputs are as follows: 0/0, 1/0, and 0/1. This means that FF #1 will have an output of 1, FF #2 will have an output of 1, and FF #3 will have an output of 0. Thus the result is 110.

THE CORRECT ANSWER IS: (D)

517. The state transitions for the circuit are 001, 100, 010, 101, 110, 111, 011, …

THE CORRECT ANSWER IS: (C)

518. One-hot encoding means that only one state bit is at logic-1 for each state transition. This will require 16 registers.

THE CORRECT ANSWER IS: (C)

COMPUTERS AFTERNOON SAMPLE SOLUTIONS

519. The instruction is $(01110100)_2$. The mode is $(110)_2$, or IN1 AND IN2 or IN1 AND ACC. The register selected is $(10)_2$, which is register 2 that has initial value $(80)_{16}$. ANDing this with the accumulator value of $(FF)_{16}$ yields a result of $(80)_{16}$.

THE CORRECT ANSWER IS: (C)

520. The instruction is $(AB6)_{16}$ or $(101010110110)_2$. Extracting the opcode yields $(0101)_2$, which corresponds to **add**.

THE CORRECT ANSWER IS: (B)

521. Each word consumes 12 bits in this architecture. The address field is 7 bits wide [6:0], thus yielding 2^7 addressable words.

THE CORRECT ANSWER IS: (B)

522. Address A8 controls the $A/\overline{B}$ select of the A/D converter. B is selected if A8=0 and $A[15:13]=(101)_2$.

THE CORRECT ANSWER IS: (D)

523. Even parity means that the number of set bits is EVEN. Response C satisfies this requirement.

THE CORRECT ANSWER IS: (C)

524. The state of C[0] for each instruction is given as: ...011100...

THE CORRECT ANSWER IS: (A)

525. The sum of the time distribution percentages for the CALL/RET/PUSH/POP instructions is $6 + 5 + 7 + 3 = 21\%$.

THE CORRECT ANSWER IS: (D)

526. The polynomial generator will always have unity as a constant term. This eliminates response (D). The feedback path is at bits 4, 5, and 8, making (C) the correct polynomial generator.

THE CORRECT ANSWER IS: (C)

527. $I_3 \oplus I_2 \oplus I_1 \oplus P_2 = 0 \Rightarrow P_2 = 1$
$I_3 \oplus I_2 \oplus I_0 \oplus P_1 = 0 \Rightarrow P_1 = 0$
$I_3 \oplus I_1 \oplus I_0 \oplus P_0 = 0 \Rightarrow P_0 = 0$

Thus, the code word is $(1111000)_2$

THE CORRECT ANSWER IS: (B)

528. **THE CORRECT ANSWER IS: (B)**

529. Using this form the following occurs:

E1: 440/200 yields a quotient of 2 and remainder of 40 (r = 40)
E2: r is not equal to 0 proceed to E3
E3: m = 200; n = 40
E1: 200/40 yields a quotient of 5 and remainder of 0 (r = 0)
E2: r = 0; terminate with 40 being the GCD

Therefore, **two** levels of recursion were required for m = 440 and n = 200.

THE CORRECT ANSWER IS: (B)

530. A compiler will flag compile-time errors for easy identification and fix.

THE CORRECT ANSWER IS: (B)

531. Develop test cases, interface tests, unit testing, regression testing.

THE CORRECT ANSWER IS: (B)

532. By necessity, the model is changed to reflect the real-world event.

THE CORRECT ANSWER IS: (C)

533. The pipeline single clock cycle must be slow enough to accommodate the slowest pipeline stage latency.

THE CORRECT ANSWER IS: (B)

534. **THE CORRECT ANSWER IS: (D)**

COMPUTERS AFTERNOON SAMPLE SOLUTIONS

535.   **THE CORRECT ANSWER IS: (B)**

536.   Arguments are pushed and then the operators are applied.

      THE CORRECT ANSWER IS: (B)

537.   The diameter of a graph is defined as the maximum length between the shortest path of any two vertices (or in this case computers).

      THE CORRECT ANSWER IS: (C)

538.   The transport layer is where the TCP protocol lies. This is the host-to-host layer where communication apparently occurs between ports on various hosts.

      THE CORRECT ANSWER IS: (D)

539.   The OSI model uses the following layers:
7	Application/user interface
6	Presentation/session management
5	Session/application transmission
4	Transport/end-to-end transmission
3	Network/pathway transmission
2	Data Link/message formulation, headers, point-to-point transmission
1	Physical/transmission over physical medium.

      THE CORRECT ANSWER IS: (C)

540.   The test set impedance is 50 Ω, but it is in series with 47 kΩ, so this has little impact on the receiver impedance.

      THE CORRECT ANSWER IS: (C)

ELECTRONICS, CONTROLS, AND COMMUNICATIONS
AFTERNOON SAMPLE SOLUTIONS

CORRECT ANSWERS TO THE ELECTRONICS, CONTROLS, AND COMMUNICATIONS AFTERNOON SAMPLE QUESTIONS

Detailed solutions for each question begin on the next page.

501	C	**521**	A
502	D	**522**	C
503	D	**523**	A
504	C	**524**	B
505	A	**525**	A
506	C	**526**	D
507	D	**527**	B
508	B	**528**	D
509	A	**529**	C
510	A	**530**	C
511	B	**531**	D
512	C	**532**	B
513	C	**533**	C
514	A	**534**	A
515	A	**535**	C
516	D	**536**	D
517	C	**537**	C
518	D	**538**	B
519	C	**539**	A
520	A	**540**	C

501.

$$V_o = -\frac{R_2}{R_1 + R_s} V_s$$

$$V_o\big|_{max} = -\frac{R_2(max)}{R_1 + R_s}(-100 \text{ mV})$$

$$= -\frac{(1.01)10^6}{10 \times 10^3}(-100 \times 10^{-3})$$

$$= 10.1$$

THE CORRECT ANSWER IS: (C)

502. The coefficients, b_i, are linear in all cases except D.

THE CORRECT ANSWER IS: (D)

503. During idle time the RXD line is at the mark level of -12 V.

THE CORRECT ANSWER IS: (D)

504. If any input is high, the input voltage to the last inverter is low, making F high and resulting in an "OR".

THE CORRECT ANSWER IS: (C)

505. General form of bandpass filter is

$$\frac{a_1 s}{s^2 + \left(\dfrac{\omega_o}{Q}\right)s + (\omega_o)^2}$$

where ω_o = center frequency

$$BW = \frac{\omega_o}{Q} = \text{3-db bandwidth}$$

10^5 rad/s

THE CORRECT ANSWER IS: (A)

506. $V_i = 4$ V, $R_i = 2$ kΩ, $R_c = 1$ kΩ
$V_+ = V_- = V_i$ and $I_+ = I_- = 0$

Then $I_c = \dfrac{V_i}{R_c} = \dfrac{4}{1\text{k}} = 4$ mA

$V_o = V_c + V_- = 2 + 4 = 6$ V

THE CORRECT ANSWER IS: (C)

507. Pins 10 (GND) and 20 (V_{cc}) are the pins associated with the power supply. Each pin has an inductance of 10.97 nH. Therefore, the effective power supply inductance is

$L_{eff} = 2(10.97 \text{ nH}) = 21.94$ nH

THE CORRECT ANSWER IS: (D)

508. $I_{B3} = I_{B4} = \dfrac{I_{C3}}{\beta}$

Total current through R_{C3} is $I_{C3} + I_{B3} + I_{B4} = \dfrac{20 - V_{BE3}}{R_{C3}}$

$I_{C3}(1 + 2/\beta) = \dfrac{20 - V_{BE3}}{R_{C3}}$

$\therefore \quad I_{C3} = \dfrac{20 - V_{BE3}}{(1 + 2/\beta)R_{C3}}$

THE CORRECT ANSWER IS: (B)

509. The current distribution in (A) results in a radiation pattern with two main lobes. The remaining distributions have radiation patterns with many more lobes.

THE CORRECT ANSWER IS: (A)

510. Free space loss $= 20 \log \left[\dfrac{4\pi d}{c/f} \right]$

$d = 8$ miles $= 12{,}874.752$ m
$f = 23.6 \times 10^9$ Hz
$c = 3 \times 10^8$ m/sec
FSL $= 141.98$ dB

THE CORRECT ANSWER IS: (A)

511. Since, by definition, the voltage is forced to zero at a short circuit, the reflected voltage will be equal to the incident voltage. With the line loss, the incident voltage will be 70.7% of the voltage at the input, and on arriving back at the input where the VSWR is measured, 70.7% of that, or 1/2 of the transmitted voltage, is present.

So, $\dfrac{|V_{max}|}{|V_{min}|} = \dfrac{1 + \dfrac{1}{2}}{1 - \dfrac{1}{2}} = 3$

THE CORRECT ANSWER IS: (B)

512. Note that the voltage drop across the series regulator is $14 - 5 = 9$ V. The current passing through the series regulator is $5/5 = 1$ A, making the total power dissipation in the regulator $(9)(1) = 9$ W. The Case Temperature Derating Curve has the form $P_{Dmax} = 15 - 0.25T$, where T is the free air temperature. Solving for T given a power dissipation of 9 W gives a maximum operating temperature of 114°C.

THE CORRECT ANSWER IS: (C)

513. At 90°, the following transistors must be conducting: a, d, and f. The control word is then (100101).

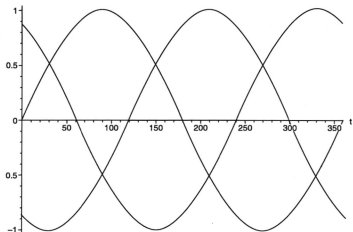

THE CORRECT ANSWER IS: (C)

514. In the two-inverter circuit the total loop phase shift must be 360° in order to sustain oscillation. At the series-resonant frequency f_s, the impedance of the crystal is only resistance. So series resonance must be used.

THE CORRECT ANSWER IS: (A)

515. The effective Thevenin base resistance is

$$R_B = R_1 \| R_2 = \frac{R_1 R_2}{R_1 + R_2} = \frac{(2 \times 10^6)(4 \times 10^6)}{2 \times 10^6 + 4 \times 10^6} = 1.33 \text{ M}\Omega$$

The Thevenin base voltage is

$$V_B = \left(\frac{R_2}{R_1 + R_2} \right)(15 \text{ V}) = \left(\frac{4 \times 10^6}{2 \times 10^6 + 4 \times 10^6} \right)(15 \text{ V}) = 10.0 \text{ V}$$

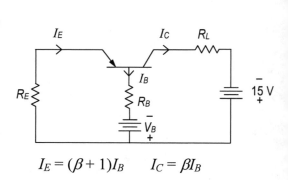

$$0 = R_E I_E + V_{EB} + R_B I_B - V_B$$
$$0 = (\beta + 1)R_E I_B + V_{EB} + R_B I_B - V_B$$
$$I_B = \frac{V_B - V_{EB}}{(\beta + 1)R_E + R_B}$$
$$I_C = \beta I_B = \frac{\beta(V_B - V_{EB})}{(\beta + 1)R_E + R_B} = \frac{(100)(10 - 0.7)}{(101)(9.9 \times 10^3) + 1.33 \times 10^6}$$
$$I_C = 0.3986 \text{ MA}$$

$$I_E = (\beta + 1)I_B \qquad I_C = \beta I_B$$

THE CORRECT ANSWER IS: (A)

168

516. $V_{gs} = V_{in}$

$$0 = \frac{V_{out} - V_{in}}{R_G} + g_m V_{in} + \frac{V_{out}}{R_D} + \frac{V_{out}}{R_L}$$

$$\left(\frac{1}{R_G} - g_m \right) V_{in} = \left(\frac{1}{R_G} + \frac{1}{R_D} + \frac{1}{R_L} \right) V_{out}$$

$$\frac{V_{out}}{V_{in}} = \frac{\dfrac{1}{R_G} - g_m}{\dfrac{1}{R_G} + \dfrac{1}{R_D} + \dfrac{1}{R_L}} = \frac{1 - g_m R_G}{1 + \dfrac{R_G}{R_D} + \dfrac{R_G}{R_L}}$$

THE CORRECT ANSWER IS: (D)

517. $V_p = \dfrac{R_2}{R_1 + R_2} V_1$

$V_n = V_p$

$$0 = \frac{V_n - V_2}{R_3} + \frac{V_n - V_{out}}{R_4}$$

$$V_{out} = \left(1 + \frac{R_4}{R_3} \right) V_n - \frac{R_4}{R_3} V_2$$

$$= \left(\frac{R_2}{R_1 + R_2} \right)\left(1 + \frac{R_4}{R_3} \right) V_1 - \frac{R_4}{R_3} V_2$$

$$V_{out} = K(V_1 - V_2)$$

$$\left(\frac{R_2}{R_1 + R_2} \right)\left(1 + \frac{R_4}{R_3} \right) = \frac{R_4}{R_3}$$

$$\frac{R_3}{R_4}\left(1 + \frac{R_4}{R_3} \right) = \frac{R_1 + R_2}{R_2}$$

$$\left(\frac{R_3}{R_4} + 1 \right) = \left(\frac{R_1}{R_2} + 1 \right)$$

$$\frac{R_3}{R_4} = \frac{R_1}{R_2}$$

THE CORRECT ANSWER IS: (C)

518. $I_Z = K[V_{GS2} - V_T]^2$

$$V_{GS2} = \sqrt{\frac{I_Z}{K}} + V_T = 1.4472 \text{ V}$$

$$V_{S2} = V_X - V_{GS2} = 3.553 \text{ V}$$

$$R = \frac{V_{S2}}{I_Z} = 355 \text{ k}\Omega$$

THE CORRECT ANSWER IS: (D)

519. $\dfrac{1}{z^{-1}(1 - 0.7z^{-1})} = \dfrac{z}{(z - 0.7)/z} = \dfrac{z^2}{z - 0.7} \Rightarrow$ pole at $z = 0.7$

THE CORRECT ANSWER IS: (C)

520. $F(s) = \dfrac{Y(s)}{X(s)} = \dfrac{K\dfrac{(s+20)}{s^2 + 2s + 2}}{1 + K\left(\dfrac{s+20}{s^2 + 2s + 2}\right)\left[\dfrac{s+150}{(s+10)(s+50)(s+100)}\right]}$

$$= \left[\frac{K(s+20)(s+10)(s+50)(s+100)}{(s^2 + 2s + 2)(s+10)(s+50)(s+100) + K(s+20)(s+150)}\right]$$

$$Y(s) = \left[\frac{K(s+20)(s+10)(s+50)(s+100)}{(s^2 + 2s + 2)(s+10)(s+50)(s+100) + K(s+20)(s+150)}\right]X(s) \qquad X(s) = 1/s$$

$$y(t)\big|_{t=0} = \text{Lim}_{s \to \infty} sY(s) \ X(s)$$
$$= \text{Lim}_{s \to \infty} sY(s)1/s = \text{Lim}_{s \to \infty} Y(s)$$

$$\text{Lim}_{s \to \infty} Y(s) = \frac{K(s^4 + \ldots\ldots)}{(s^5 + \ldots\ldots) + K(s^2)}$$

$$\text{Lim}_{s \to \infty} F(s) = \frac{K}{s} = 0$$

THE CORRECT ANSWER IS: (A)

521. $\dfrac{Y}{R} = G_4 + \dfrac{G_1 G_2}{1 + G_1 H_2}$

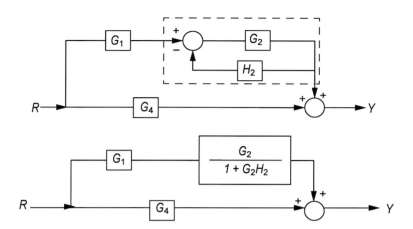

THE CORRECT ANSWER IS: (A)

522. Collapsing the feedback loop yields the following transfer function:

$$\frac{Y(s)}{U(s)} = \frac{\dfrac{1}{s^2 + 3s + 4}}{\left[1 + \dfrac{2}{s^2 + 3s + 4}\right]}(s + 4) = \frac{s + 4}{s^2 + 3s + 6}$$

Breaking this into two functions, $\dfrac{X(s)}{U(s)} \dfrac{Y(s)}{X(s)}$, yields:

$$\frac{X(s)}{U(s)} = \frac{1}{s^2 + 3s + 6} \quad \text{and} \quad \frac{Y(s)}{X(s)} = s + 4$$

This yields the following differential equations:

$$\ddot{x} = u - 3\dot{x} - 6x \quad \text{and} \quad y = \dot{x} + 4x$$

Substituting $x(t) = x_1(t)$ and $dx(t)/dt = x_2(t)$.

$$\begin{bmatrix} \dot{x}_1 \\ \dot{x}_2 \end{bmatrix} = \begin{bmatrix} 0 & 1 \\ -6 & -3 \end{bmatrix} \begin{bmatrix} x_1 \\ x_2 \end{bmatrix} + \begin{bmatrix} 0 \\ 1 \end{bmatrix} u \quad \text{and} \quad y = \begin{bmatrix} 4 & 1 \end{bmatrix} \begin{bmatrix} x_1 \\ x_2 \end{bmatrix}$$

THE CORRECT ANSWER IS: (C)

171

523. $G_c = K \dfrac{(s + Z_1)}{s + P_1} = \dfrac{K(s + 50)}{s + 200}$

$P_1 > Z_1$ phase lead

THE CORRECT ANSWER IS: (A)

524. Lead-lag or lag-lead compensators have the general form:

$$K \left[\frac{s + \beta}{s + \gamma} \right]$$

The PI controller has the form

$$G_c(s) = A_p + \frac{A_i}{s} = K \left[\frac{s + a}{s} \right]$$

Thus the PI controller is near in form to a L/L compensator.

THE CORRECT ANSWER IS: (B)

525. The plant transfer function has poles at $s = 0$, -1, and -5. The root locus shows poles at $s = 0$, -1, -5, and -10 along with a zero at $s = -1$. The zero and pole at $s = -1$ cancel in the closed-loop transfer function. The compensator adds a zero at $s = -1$ and a pole at $s = -10$. The compensator therefore has a transfer function of

$$H(s) = A \left(\frac{s + 1}{s + 10} \right)$$

THE CORRECT ANSWER IS: (A)

526. Specification is value of K needed for 60° phase margin.

The phase $= -180° + 60° = -120°$ on figure

This corresponds to -21 dB

$\therefore K$ can be increased to make open-loop gain

At $-120°$ phase shift $= 1$, or K can be increased by 21 dB

$\therefore 20 \log_{10} K = 21$

$\log_{10} K = 21/20 = 1.05$

$K = 11.22$

THE CORRECT ANSWER IS: (D)

527. The initial slope of the magnitude plot is –20 db/dec, and the initial phase shift of the phase plot is –90°. Either of these facts requires that there be a pole at the origin. That is, there is a $1/s$ term in the open-loop transfer function.

The final slope of the magnitude plot is –60 db/dec, and the final phase shift of the phase plot is –270°. Either of these facts requires that the pole-zero excess be three. This requires at least two poles other than the one at the origin.

In addition, the response has a resonance peak at the breakpoint at 100 Hz. This requires a complex pole pair.

THE CORRECT ANSWER IS: (B)

528. The gain margin is the difference between the actual gain and zero dB at the frequency at which the open-loop phase shift is –180°. By inspection of the frequency plots, the frequency at which the phase shift is –180° is approximately 300 rad/s. This leads to a gain margin of approximately 56 dB.

THE CORRECT ANSWER IS: (D)

529. $x(t) = 40 \cos (200k \pi t) + 10 \cos (197k \pi t) + 10 \cos (203k \pi t)$
$x(t) = 40 [\cos (200k \pi t) + 0.25 \cos (197k \pi t) + 0.25 \cos (203k \pi t)]$

$$\frac{\max\left[\frac{1}{2}m(t)\right]}{A_c} = 0.25$$

$$\max[m(t)] = 20$$

$$\text{Mod} = \frac{2\max\left[\frac{1}{2}m(t)\right]}{A_c} = 0.5$$

$$\% \text{ mod} = 100 \times 0.5 = 50\%$$

THE CORRECT ANSWER IS: (C)

530. The modulation index for an FM signal is defined as

$$\beta_f = \frac{\Delta F}{B}$$

where ΔF is the peak frequency deviation (in this case stated to be 64 Hz) divided by B. The bandwidth B of the modulating signal for a sinusoid is f_m.

$m(t) = 8 \cos (8\pi t) = 8 \cos (\omega_m t)$
$\therefore 8\pi t = \omega_m t$
$\omega_m = 8\pi = 2\pi \times 4 \text{ Hz} = 2\pi f_m$

thus $\beta_f = \dfrac{64}{4} = 16$

THE CORRECT ANSWER IS: (C)

531. $B_T = 2(\beta + 1) B_m$ where
 B_m = Bandwidth of modulating signal
 β = Modulation index
 B_T = Bandwidth of FM signal

$B_m = \dfrac{10^7}{2\pi}$ Hz
$\beta = 10$
$B_T = 3.50 \times 10^7$ Hz

THE CORRECT ANSWER IS: (D)

532. The information content of an AM waveform is contained in the magnitude envelope of the carrier signal. Consequently, removing the carrier to leave the envelope yields the desired demodulated signal.

THE CORRECT ANSWER IS: (B)

533. $\omega_i(t) = \dfrac{d}{dt}\left[6.2\times10^{11}t + 10\cos\left(10^7 t\right)\right]$

$\omega_i(t) = 6.2\times10^{11} - 10^8\cos\left(10^7 t\right)$

$\omega_d(t) = -10^8\cos(10^7 t)$

$\Delta f = \dfrac{1}{2\pi}\max\left[\omega_d(t)\right] = 1.59\times10^7 \text{ Hz}$

THE CORRECT ANSWER IS: (C)

534. f_s = sampling frequency

f_m = maximum signal frequency

$f_m \le 1/2\, f_s$

THE CORRECT ANSWER IS: (A)

535. $F_1 = 5 \text{ dB} = 3.16$ $F_2 = 15 \text{ dB} = 31.62$ $F_3 = 25 \text{ dB} = 316.23$

$G_1 = 10 \text{ dB} = 10$ $G_2 = 15 \text{ dB} = 31.62$ $G_3 = 20 \text{ dB} = 100$

$F_T = F_1 + \dfrac{F_2 - 1}{G_1} + \dfrac{F_3 - 1}{G_1 G_2} = 7.221 = 8.586 \text{ dB}$

THE CORRECT ANSWER IS: (C)

536. $A_c = 1 \text{ m}^2$

$P_R = -40 \text{ dBm} + 10 \text{ dB} = -30 \text{ dBm}$

$= 10^{-6} \text{ W}$

$\dfrac{E_{\text{rms}}^2}{377\,\Omega} = \dfrac{P_R}{A_c} = 10^{-6} \text{ W/m}^2$

$E_{\text{rms}} = \sqrt{(377)(10^{-6})} = 0.0194 \text{ V/m}$

$= 19.4 \text{ mV/m}$

THE CORRECT ANSWER IS: (D)

537. $E_{rms} = \sqrt{kTR\Delta f}$

$$1.5 \times 10^{-6} = \sqrt{(1.375)(10^{-23})(50)(10^7)T}$$

$$T = \frac{2.25 \times 10^{-12}}{(1.375)(10^{-23})(50)(10^7)} = 327\,K$$

THE CORRECT ANSWER IS: (C)

538. The actual velocity is:

$$v_p = \frac{2L}{\tau_0}$$

The ratio of the actual velocity to the velocity of the light is therefore:

$$\frac{v_p}{v_c} = \frac{2L}{\tau_0 v_c}$$

where: $L = 1,000\,M$
$\tau_0 = 9.43\,\mu s$
$v_c = 3 \times 10^8\,M/s$

The ratio is therefore equal to 0.70696.

THE CORRECT ANSWER IS: (B)

539. ATM provides connection-oriented service with both permanent and switched virtual circuits.

THE CORRECT ANSWER IS: (A)

540. Since $P \propto v^2$

$$|\Gamma| = \frac{|v_{ref}|}{|v_{inc}|} = \sqrt{\frac{P_{ref}}{P_{inc}}} = \sqrt{\frac{1}{5}} = 0.447$$

THE CORRECT ANSWER IS: (C)

POWER
AFTERNOON SAMPLE SOLUTIONS

CORRECT ANSWERS TO THE POWER AFTERNOON SAMPLE QUESTIONS

Detailed solutions for each question begin on the next page.

501	C	**521**	B
502	A	**522**	C
503	A	**523**	D
504	C	**524**	D
505	A	**525**	B
506	D	**526**	A
507	B	**527**	D
508	D	**528**	B
509	C	**529**	B
510	C	**530**	B
511	C	**531**	A
512	D	**532**	D
513	A	**533**	C
514	D	**534**	A
515	B	**535**	C
516	B	**536**	D
517	C	**537**	C
518	A	**538**	A
519	D	**539**	B
520	B	**540**	A

POWER AFTERNOON SAMPLE SOLUTIONS

501. $I_{primary}$ = 5 A × (400 turns/5 turns) = 400 A

THE CORRECT ANSWER IS: (C)

502. The circuit would need to be balanced out to account for the 30° phase shift between wye and delta windings and, also, for the turns ratio of the transformer. Otherwise, there would be a non-zero current through the relay restraint windings during non-fault conditions. Answer (A).

The idea that CTs must be physically wired to maintain some arbitrary terminal nomenclature seems nonsensical; thus, eliminate (B).

The American standard is, indeed, to label the high-side terminal H_1 and the low-side terminal X_1 such that voltage H_1-neutral leads X_1-neutral by 30°, and the phase shift will be 30°, not more than 30°. But no matter how the CTs are connected, it will not affect the phasing of the transformer windings themselves; thus, eliminate (C).

There is no reason why a delta winding must be on either the low or high side of a transformer; thus, eliminate (D).

THE CORRECT ANSWER IS: (A)

503. Even though the high side of the transformer is connected delta (ungrounded), a strike to a 12-kV phase conductor will impose a high voltage to ground, which is best dissipated by surge arresters connected between the phases and ground. Answer (A).

Phase-to-phase arrester location will not be as sensitive to ground currents caused by lightning strikes as a phase-to-ground location. This eliminates (B).

The problem statement refers to protection against lightning strikes on the 12-kV line. Arresters on the 480-V side would not protect the transformer against such an event. Eliminate (C) and (D).

THE CORRECT ANSWER IS: (A)

504. 10-hp load is 50 A × 1.25 = 62.5 A (***NEC*** Table 430.148, ***NEC*** Article 430.22 (A)

$$\text{Resistance heating load} = \frac{1,000 \text{ W}}{240 \text{ V}} = 4.2 \text{ A}$$

Total load = 62.5 + 4.2 = 66.7 A

Select #4 copper (***NEC*** Table 310.16)

THE CORRECT ANSWER IS: (C)

505. Refer to *NEC*® Table 430.151(B) for the locked-rotor current applicable to this motor, 412 A. (Locked-rotor current is usually considered the same as starting current.)

THE CORRECT ANSWER IS: (A)

506. *NEC*® Article 430.52(c) specifies that motor circuit conductors be protected against short circuit or ground faults in accordance with Table 430.52. Answer (D).

The ampacity of the conductors is not a factor in sizing these fuses since the fuses protect only against shorts and ground faults. The conductors are protected against overload by the motor-overload device. Eliminate (A).

The conductors themselves (but not the fuses) are usually sized based upon 125% of full-load motor amps per Article 430.22(A). Eliminate (B).

Motor locked-rotor current is related to the required interrupting capacity of motor disconnect switches (per Article 430.110), but is not related to fuses for conductor protection. Eliminate (C).

THE CORRECT ANSWER IS: (D)

507. Base impedance = $(kV_{base})^2 / (MVA_{base})$, where kV_{base} is understood to be the phase-to-phase voltage and MVA_{base} is the 3-phase MVA.

$= (65 \text{ kV})^2/(24 \text{ MVA}) = 176 \ \Omega$

pu impedance = 50 Ω/176 Ω = 0.284

THE CORRECT ANSWER IS: (B)

508. For this unbalanced load, $I_A + I_B + I_N = 0$
Also, $V_{\phi N} = V_{\phi\phi} / \sqrt{3} = 13.2 / 1.732 = 7.62$ kV

$$\left| I_A + I_B \right| = \left| -I_N \right| = \left| \frac{200 + j100}{7.62\angle 0°} + \frac{200 + j100}{7.62\angle 120°} \right| = 29.3 \text{ A}$$

THE CORRECT ANSWER IS: (D)

509. The system is initially ungrounded (the utility neutral is disconnected from ground). Connecting Corner A of the delta to ground will therefore have no effect on the relative phase voltages and $V_{BG} = V_{BA} = 13.2$ kV.

THE CORRECT ANSWER IS: (C)

POWER AFTERNOON SAMPLE SOLUTIONS

510. $I_C = \dfrac{500 \text{ kVA}}{13.2 \text{ kV}} = 37.9 \text{ A}$

THE CORRECT ANSWER IS: (C)

511. Since the distribution system is balanced,

$$V_{an} = \frac{12.5}{\sqrt{3}} \angle -30° + \frac{70 \angle -20°}{1,000}(5 + j10) = 7.48 \angle -24.2°$$

$$|V_{ab}| = 7.48\sqrt{3} = 12.95 \text{ kV}$$

THE CORRECT ANSWER IS: (C)

512. Positive-sequence and negative-sequence currents cannot pass through the neutral grounding resistor in a wye-connected generator. Answer (D).

The three zero-sequence current phasors are equal in magnitude, but are all in phase; therefore eliminate (A).

The positive- and negative-sequence magnitudes are not necessarily the same; therefore, eliminate (B).

The voltage drop through the neutral grounding resistor for a wye-connected synchronous generator is $3I_0R$ (since all three zero-sequence currents are in phase, they are additive); therefore, eliminate (C).

THE CORRECT ANSWER IS: (D)

513. Since the power supply is loaded, there must be nonzero charging current being supplied to the capacitor. It is reasonable to assume steady-state conditions. It is also reasonable to assume that the forward voltage drop across the diodes is negligible compared to 120 V and that the conductance of the diodes is zero when reverse biased.

Sketch the current for one ac phase. The single-phase neutral current will consist of one positive pulse and one negative pulse, per 1/60-second cycle, when the ac voltage exceeds the dc voltage on the filter capacitor and the capacitor is charging. Each of the other two ac phases will contribute similar current pulses on the neutral, shifted by 120° and 240° (1/180 and 2/180 seconds), respectively. The pulses will not cancel; the primary frequency on the neutral will be the first triplen, i.e., 180 Hz. (Other, higher, triplen harmonics will also be present.) Answer (A).

(B) shows the correct number of current pulses per cycle, but they do not alternate, therefore, eliminate (B).

POWER AFTERNOON SAMPLE SOLUTIONS

(C) shows only three pulses per cycle, each of the same polarity, therefore, eliminate (C).

(D) a flat trace, i.e., zero neutral current, would occur if there were no charging current being supplied to the capacitor; since the power supplies are resistively loaded, the charging current must be nonzero, therefore, eliminate (D).

THE CORRECT ANSWER IS: (A)

514. This inverter essentially produces 3-phase square waves by sequentially switching the transistors on and off to drive the motor. For the purpose of this question, it is reasonable to neglect the forward voltage drop across the transistors when they are conducting.

Pick any two transistors that are connected to opposite sides of the dc link and to different motor phases. Assume these two transistors are turned on. The phase-to-phase voltage across those two motor phases is equal to the dc-link voltage.

THE CORRECT ANSWER IS: (D)

515. $I_{AVE} = \dfrac{1}{T} \displaystyle\int_0^T i(t)\,dt$, where T is the period of the waveform, $180°$

$$= \frac{1}{\pi} \int_{45°}^{180°} \sin(\phi)\,d\phi$$

$$= -\frac{1}{\pi} \cos(\phi) \Big|_{45°}^{180°}$$

$$= 0.543 \text{ A}$$

THE CORRECT ANSWER IS: (B)

182

516. The path length for Leg B is *d*, and the path length for Leg C is 3*d*. Thus, the flux will divide between Legs B and C in the ratio of 1:3. Therefore, 1/4 of the total flux will be in Leg C and 3/4 will be in Leg B.

$$\varphi_C = 1/4 \times 3.0 \times 10^{-3} = 0.75 \times 10^{-3} \text{ Wb}$$

THE CORRECT ANSWER IS: (B)

517. Geometric Mean Distance $= (D_{ab} \times D_{ac} \times D_{bc})^{1/3}$

$$= (3 \times 4 \times 7)^{1/3} = 4.38 \text{ ft}$$

THE CORRECT ANSWER IS: (C)

518. $I_G = \dfrac{132}{13.2} \, I_L = 10 \, I_L = 10 \times 75.93 = 759.3 \text{ A}$

THE CORRECT ANSWER IS: (A)

519. Note that only the mechanical motor properties are relevant to the question, not the electrical properties. The following equation relates motor speed, torque, speed, and shaft horsepower:

hp = torque (ft-lb) × speed (rpm) × (hp/33,000 ft-lb/min) × (2π rad/rev)

Solving for full-load torque:

$$T_{FL} = 1{,}000 \text{ hp} \times 33{,}000 \text{ ft-lb/min-hp} / (1{,}165 \text{ rpm} \times 2\pi \text{ rad/rev})$$

$$= 4{,}508 \text{ ft-lb}$$

$$T_{START} = 150\% \, T_{FL} = 1.5 \times 4{,}508 = 6{,}762 \text{ ft-lb}$$

THE CORRECT ANSWER IS: (D)

520. $\text{Reg} = (\eta_{NL} - \eta_{FL}) \times 100 / \eta_{FL} = (1{,}790 - 1{,}750) \times 100 / 1{,}750 = 2.3\%$

THE CORRECT ANSWER IS: (B)

POWER AFTERNOON SAMPLE SOLUTIONS

521. Generator $I_{out} = \dfrac{1{,}131\,\text{kVA}}{(2.4\,\text{kV})\left(\sqrt{3}\right)} = 272\,\text{A}$

THE CORRECT ANSWER IS: (B)

522. Generator Power$_{out} = (150\,\text{MVA})(0.85) = 127.5\,\text{MW}$

THE CORRECT ANSWER IS: (C)

523. At no load, the current is mostly reactive. Thus, reactive motor current at no load = 28 A and is nearly independent of motor load.

Full-Load Conditions   Half-Load Conditions

Real current at full load = 87.6 A. See full-load phasor diagram above.

At half load, the real current is 50% of 87.6 = 43.8 A, the power factor angle = 32.6°.

Power factor at half load = cos 32.6° = 0.84

THE CORRECT ANSWER IS: (D)

524. The wye-start, delta-run design feature reduces the starting voltage to the motor and hence its starting current. Answer (D).

High-efficiency motors tend to have larger wire, more iron, and thinner laminations to reduce the internal impedance of the motor. This may increase starting current. Eliminate (A).

Locked-rotor current is not a function of load; however, starting a motor under loaded conditions can increase the time it takes it to reach its operating speed and cause its start-up temperature to rise. Eliminate (B).

A higher service factor is a measure of the overload capability of the motor and is not related to starting current. Eliminate (C).

THE CORRECT ANSWER IS: (D)

525. The transformers will load in inverse proportion to their impedances.

Converting T_1 impedance to T_2 Base: $j0.045\left(\dfrac{2{,}000}{1{,}000}\right) = j0.09$

$$\frac{\text{Load}_{T_1}}{\text{Load}_{T_2}} = \frac{Z_{T_2}}{Z_{T_1}}$$

When T_1 is fully loaded to 1,000 kVA,

$$\text{Load}_{T_2} = (1{,}000)\left(\frac{0.09}{0.06}\right) = 1{,}500 \text{ kVA}$$

Total Load = 1,500 + 1,000 = 2,500 kVA

THE CORRECT ANSWER IS: (B)

526. If the transformers are paralleled, the voltages of the secondary windings will be 30° out of phase, resulting in excessive circulating currents. Changing the turn ratios of the transformers or reconfiguring the neutral connections will not affect the 30° phase shift between the secondary winding voltages. Thus transformers should not be paralleled.

THE CORRECT ANSWER IS: (A)

527. Transformer impedance on the new base $= 0.08\left(\dfrac{100\,\text{MVA}}{25\,\text{MVA}}\right) = 0.32$ pu

THE CORRECT ANSWER IS: (D)

528. A 3-phase induction motor in which the inverter allows the voltage-to-frequency ratio to increase may saturate. Answer (B).

Operating a 50-Hz transformer at higher than rated frequency will reduce the chances of saturating the transformer. Eliminate (A).

Geomagnetic storms do not cause saturation in delta transformers since the delta is not referenced to ground. Eliminate (C).

The normal current transformer configuration is with a short-circuited secondary winding. Opening the secondary circuit could cause saturation. Eliminate (D).

THE CORRECT ANSWER IS: (B)

529. Feeder impedance $= \left(\dfrac{250}{1,000}\right)(0.029 + j0.048) = 0.00725 + j0.0120$ (*NEC®* Chapter 9, Table 9)

$$V_{Panel\ A} = \left(\dfrac{480}{\sqrt{3}}\right)\angle 0° - \left(400\angle - \cos^{-1} 0.80\right)(0.00725 + j0.0120) = 271.94\angle -0.44°$$

$$\left|V_{Phase\ to\ Phase}\right| = 271.94 \times \sqrt{3} = 471\ \text{V}$$

An alternate solution may use the approximation for effective Z given in Note 2, Chapter 9, Table 9 of the *NEC®* . This approach will also result in Answer (B).

THE CORRECT ANSWER IS: (B)

530. Since, after addition of the capacitor, the motor current will be more than the current seen by the overload relays, the relay trip settings should be reduced (*NEC®* Article 460.9). Answer (B).

The current at the motor will not change since the capacitor will reduce current only between itself and the source. Eliminate (A).

A separate disconnect switch is not required when the capacitor is connected on the load side of a motor controller [*NEC®* Article 460.8 (c) exception]. Eliminate (C).

The installation of a capacitor does not change the required motor feed conductor size (*NEC®* Article 460.9). Eliminate (D).

THE CORRECT ANSWER IS: (B)

531. The vars vary as the square of the applied voltage.

$$\text{kvar} = \left(\dfrac{208}{240}\right)^2 (110) = 82.6$$

THE CORRECT ANSWER IS: (A)

532. $I_{\text{Fault}} = \dfrac{60,000/\sqrt{3}}{16.75\angle 71° + 13.4\angle 71° + 13.25\angle 81°} = 800\ \text{A}$

THE CORRECT ANSWER IS: (D)

533. *NEC®* Article 250.56

THE CORRECT ANSWER IS: (C)

POWER AFTERNOON SAMPLE SOLUTIONS

534. The grounding resistor at the neutral of the transformer will limit ground-fault current. Answer (A).

For balanced loads, the potential difference between N and EG is 0. Eliminate (B).

The interrupting rating of M is sized for 3-phase faults. Eliminate (C).

THE CORRECT ANSWER IS: (A)

535. Solution should use the transient reactance, which is 0.25.

$$E_0' = E_t + jX_d'(I_t) = 1.0 + j(0.25)(1.0 + j0) = 1.03 \text{ pu}$$

THE CORRECT ANSWER IS: (C)

536. The voltage drop will be largest for a lagging power factor of 0.707.

The approximate voltage drop magnitude is given by: $V = I(X\sin\phi + R\cos\phi)$, where ϕ is the power factor angle (see Chapter 9, Table 9 of the **NEC®** Handbook).

Since $X = R$, voltage drop magnitude $\propto (\sin\phi + \cos\phi)$

At 0.707 power factor, $(\sin\phi + \cos\phi) = 0.707 + 0.707 = 1.414$
At unity power factor, $\phi = 0°$ and $(\sin\phi + \cos\phi) = 1.0$
At zero power factor, $\phi = 90°$ and $(\sin\phi + \cos\phi) = 1.0$

THE CORRECT ANSWER IS: (D)

537. Multiple of pick-up setting $= \dfrac{\text{Current}}{\text{Pick-up}} = \dfrac{1,500}{420} = 3.6\times$

The time dial setting for $3.6\times$ at 0.9 sec $= 2$

THE CORRECT ANSWER IS: (C)

538. For "normal" applications that do not qualify for the listed exceptions, overcurrent protection shall not exceed 30 A. **NEC®** 240.3 (D)

THE CORRECT ANSWER IS: (A)

539. Current during the maximum fault condition $= 8{,}000 \left(\dfrac{5}{400} \right) = 100$ A

Excitation voltage $= (100 \text{ A})(1.1 \ \Omega) = 110$ V

THE CORRECT ANSWER IS: (B)

540. Required $V_{NG} = \dfrac{120 \times 5}{1{,}000} = 0.6$ kV

THE CORRECT ANSWER IS: (A)

APPENDIX A

SAMPLE OF EXAM COVERS AND INSTRUCTIONS

NAME: _____

Last   First   Middle Initial

ELECTRICAL AND COMPUTER SAMPLE–AM

PRINCIPLES AND PRACTICE OF ENGINEERING EXAMINATION

Answers to multiple-choice questions must be placed on the separate answer sheet

SAMPLE–AM

SAMPLE

NATIONAL COUNCIL OF EXAMINERS FOR ENGINEERING AND SURVEYING

ELECTRICAL AND COMPUTER ENGINEERING SAMPLE

Morning Session

THIS IS AN OPEN-BOOK EXAMINATION

NOTE: LOCAL LICENSURE BOARD RULES TAKE PRECEDENCE OVER THE FOLLOWING INSTRUCTIONS

Textbooks, handbooks, <u>bound</u> reference materials and battery-operated, silent, non-printing calculators may be used in the examination room. Writing tablets, unbound tables or notes, and devices that may compromise the security of the examination are **NOT PERMITTED** in the examination room. The exchange of reference materials during the examination is not allowed. This is an examination of your capabilities; the work is to be representative of your knowledge. Copying or cheating of any kind is **NOT** tolerated and will result in an invalidation of your examination score.

The examination is a maximum of four hours in length. Work all **40** questions according to the proctor's instructions. Each question is worth one point for the correct response; points are not subtracted for incorrect responses. The maximum possible score for this part of the examination is 40 points.

Be sure to correctly enter all of your answers on the separate answer sheet enclosed within this examination booklet. **THIS IS THE ONLY RECORD OF THE QUESTIONS YOU HAVE WORKED.** Once you select your answer, fill in the corresponding space on the answer sheet using a Number 2 pencil. **BE SURE THAT EACH MARK IS DARK AND COMPLETELY FILLS THE ANSWER SPACE.** Only one answer is permitted for each question; no credit is given for multiple answers. If you change an answer, be sure to completely erase the previous mark. Incomplete erasures may be read as intended answers.

Blank space in the examination booklet may be used for scratch work. **NO CREDIT WILL BE GIVEN FOR ANY WORK WRITTEN IN THE EXAMINATION BOOKLET.**

Make certain that you have followed the above instructions before turning in all examination materials to the proctor.

YOU ARE PROHIBITED FROM COPYING THE QUESTIONS FOR FUTURE REFERENCE;

VIOLATORS WILL BE PROSECUTED TO THE FULL EXTENT OF THE LAW.

SAMPLE

NAME: _____
Last   First   Middle Initial

National Council of Examiners for Engineering and Surveying

ELECTRICAL SAMPLE–PM

PRINCIPLES AND PRACTICE OF ENGINEERING EXAMINATION

Answers to multiple-choice questions must be placed on the separate answer sheet

SAMPLE–PM

SAMPLE

NATIONAL COUNCIL OF EXAMINERS FOR ENGINEERING AND SURVEYING

ELECTRICAL AND COMPUTER ENGINEERING SAMPLE

Afternoon Session

THIS IS AN OPEN-BOOK EXAMINATION

NOTE: LOCAL LICENSURE BOARD RULES TAKE PRECEDENCE OVER THE FOLLOWING INSTRUCTIONS

Textbooks, handbooks, <u>bound</u> reference materials and battery-operated, silent, non-printing calculators may be used in the examination room. Writing tablets, unbound tables or notes, and devices that may compromise the security of the examination are **NOT PERMITTED** in the examination room. The exchange of reference materials during the examination is not allowed. This is an examination of your capabilities; the work is to be representative of your knowledge. Copying or cheating of any kind is **NOT** tolerated and will result in an invalidation of your examination score.

This examination booklet contains three examinations in depth areas (sub-disciplines) of electrical engineering. The three examinations are listed in the table of contents at the front of this examination booklet. According to the proctor's instructions, you should select and work **one** examination of your choice. Each examination is a maximum of four hours in length and presents 40 questions. Each question is worth one point for the correct response; points are not subtracted for incorrect responses. The maximum score for this part of the examination is 40 points. Be sure to correctly indicate the examination you have worked on the separate answer sheet enclosed within this booklet. **THIS IS THE ONLY RECORD OF THE EXAMINATION YOU HAVE WORKED.**

Be sure to correctly enter all of your answers on the separate answer sheet enclosed within this examination booklet. **THIS IS THE ONLY RECORD OF THE QUESTIONS YOU HAVE WORKED.** Once you select your answer, fill in the corresponding space on the answer sheet using a Number 2 pencil. **BE SURE THAT EACH MARK IS DARK AND COMPLETELY FILLS THE ANSWER SPACE.** Only one answer is permitted for each question; no credit is given for multiple answers. If you change an answer, be sure to completely erase the previous mark. Incomplete erasures may be read as intended answers.

Blank space in the examination booklet may be used for scratch work. **NO CREDIT WILL BE GIVEN FOR ANY WORK WRITTEN IN THE EXAMINATION BOOKLET.**

Make certain that you have followed the above instructions before turning in all examination materials to the proctor.

YOU ARE PROHIBITED FROM COPYING THE QUESTIONS FOR FUTURE REFERENCE;

VIOLATORS WILL BE PROSECUTED TO THE FULL EXTENT OF THE LAW.

APPENDIX B
SAMPLE ANSWER SHEETS

A.M. **NATIONAL COUNCIL OF EXAMINERS FOR ENGINEERING & SURVEYING** (FOR NCEES USE ONLY) ▶ ○ AM ONLY **A.M.**

① LAST NAME
First 4 letters | 1st. Init.

② BOARD FROM WHICH LICENSURE IS SOUGHT

○ Ala. ○ La. ○ Okla.
○ Alaska ○ Maine ○ Oreg.
○ Ariz. ○ Md. ○ Pa.
○ Ark. ○ Mass. ○ P.R.
○ Calif. ○ Mich. ○ R.I.
○ Colo. ○ Minn. ○ S.C.
○ Conn. ○ Miss. ○ S. Dak
○ Del. ○ Mo. ○ Tenn.
○ D.C. ○ Mont. ○ Tex.
○ Fla. ○ MP   ○ Utah
○ Ga.   ○ Neb. ○ V.I.
○ Guam ○ Nev. ○ Vt.
○ Hawaii ○ N.H. ○ Va.
○ Idaho ○ N.J. ○ Wash.
○ Ill. ○ N. Mex ○ W. Va.
○ Ind. ○ N.Y. ○ Wis.
○ Iowa ○ N.C. ○ Wyo.
○ Kans. ○ N. Dak
○ Ky.   ○ Ohio

③ PRINT INSIDE BOXED-IN AREA

EXAM BOOKLET SERIAL NUMBER
(from top right corner of book)

④ DATE OF BIRTH

MONTH | DAY | 19-YEAR
○ Jan ○ Feb ○ Mar ○ Apr ○ May ○ June ○ July ○ Aug ○ Sept ○ Oct ○ Nov ○ Dec

⑤ EXAMINEE IDENTIFICATION NUMBER

INSTRUCTIONS — USE NO. 2 PENCIL ONLY
• Do NOT use ink or ballpoint pen.
• Erase completely any marks you wish to change.
• Make NO stray marks on this answer sheet.
• Incomplete erasures and stray marks may be read as intended answers.
• Make heavy black marks that completely fill the circle.
IMPROPER MARK ⊘ ⊗ ⊖ ⊙   PROPER MARK ●

⑥ PLEASE READ, THEN SIGN YOUR NAME BELOW:

I affirm by my signature below that I am the person taking this exam, the answers contained hereon are solely of my knowledge and hand, and I have not taken this exam in the previous 30 days.

I further affirm that I will not copy any information onto material to be taken from the exam room. Nor will I reveal in whole or in part any exam questions, answers, problems or solutions to anyone during or after the exam, whether orally, in writing, or any internet "chat rooms," or otherwise. I understand that failure to comply with this statement could result in an invalidation of my exam results and/or bar me from retaking the exam for a time at the discretion of the board.

Signature

⑦ Please PRINT your NAME below:

Last name   First name   Middle Initial

EXAM DATE ___ Month ___ Day ___ Year

LOCATION ___ City ___ State

DARKEN THE CIRCLE BESIDE THE DISCIPLINE TO BE WORKED. THE FOLLOWING DISCIPLINES OFFER A CHOICE OF QUESTIONS. WORK ONLY 40 QUESTIONS WHICH MUST BE ANSWERED IN 4 SETS OF 10.

○ Metallurgical
○ Mining/Mineral

DARKEN THE CIRCLE BESIDE THE DISCIPLINE TO BE WORKED. THE FOLLOWING DISCIPLINES ARE NO-CHOICE. QUESTIONS 101–140 WILL BE SCORED (FOR ENVIRONMENTAL, QUESTIONS 101–150 WILL BE SCORED.)

○ Agricultural
○ Chemical
○ Civil
○ Control Systems
○ Environmental
○ Fire Protection
○ Industrial
○ Manufacturing
○ Mechanical
○ Naval Arch./Marine
○ Nuclear
○ Petroleum
○ Structural I

DARKEN CIRCLE TO HAVE QUESTIONS 101–110 SCORED
101 Ⓐ Ⓑ Ⓒ Ⓓ
102 Ⓐ Ⓑ Ⓒ Ⓓ
103 Ⓐ Ⓑ Ⓒ Ⓓ
104 Ⓐ Ⓑ Ⓒ Ⓓ
105 Ⓐ Ⓑ Ⓒ Ⓓ
106 Ⓐ Ⓑ Ⓒ Ⓓ
107 Ⓐ Ⓑ Ⓒ Ⓓ
108 Ⓐ Ⓑ Ⓒ Ⓓ
109 Ⓐ Ⓑ Ⓒ Ⓓ
110 Ⓐ Ⓑ Ⓒ Ⓓ

DARKEN CIRCLE TO HAVE QUESTIONS 111–120 SCORED
111 Ⓐ Ⓑ Ⓒ Ⓓ
112 Ⓐ Ⓑ Ⓒ Ⓓ
113 Ⓐ Ⓑ Ⓒ Ⓓ
114 Ⓐ Ⓑ Ⓒ Ⓓ
115 Ⓐ Ⓑ Ⓒ Ⓓ
116 Ⓐ Ⓑ Ⓒ Ⓓ
117 Ⓐ Ⓑ Ⓒ Ⓓ
118 Ⓐ Ⓑ Ⓒ Ⓓ
119 Ⓐ Ⓑ Ⓒ Ⓓ
120 Ⓐ Ⓑ Ⓒ Ⓓ

DARKEN CIRCLE TO HAVE QUESTIONS 121–130 SCORED
121 Ⓐ Ⓑ Ⓒ Ⓓ
122 Ⓐ Ⓑ Ⓒ Ⓓ
123 Ⓐ Ⓑ Ⓒ Ⓓ
124 Ⓐ Ⓑ Ⓒ Ⓓ
125 Ⓐ Ⓑ Ⓒ Ⓓ
126 Ⓐ Ⓑ Ⓒ Ⓓ
127 Ⓐ Ⓑ Ⓒ Ⓓ
128 Ⓐ Ⓑ Ⓒ Ⓓ
129 Ⓐ Ⓑ Ⓒ Ⓓ
130 Ⓐ Ⓑ Ⓒ Ⓓ

DARKEN CIRCLE TO HAVE QUESTIONS 131–140 SCORED
131 Ⓐ Ⓑ Ⓒ Ⓓ
132 Ⓐ Ⓑ Ⓒ Ⓓ
133 Ⓐ Ⓑ Ⓒ Ⓓ
134 Ⓐ Ⓑ Ⓒ Ⓓ
135 Ⓐ Ⓑ Ⓒ Ⓓ
136 Ⓐ Ⓑ Ⓒ Ⓓ
137 Ⓐ Ⓑ Ⓒ Ⓓ
138 Ⓐ Ⓑ Ⓒ Ⓓ
139 Ⓐ Ⓑ Ⓒ Ⓓ
140 Ⓐ Ⓑ Ⓒ Ⓓ

DARKEN CIRCLE TO HAVE QUESTIONS 141–150 SCORED
141 Ⓐ Ⓑ Ⓒ Ⓓ
142 Ⓐ Ⓑ Ⓒ Ⓓ
143 Ⓐ Ⓑ Ⓒ Ⓓ
144 Ⓐ Ⓑ Ⓒ Ⓓ
145 Ⓐ Ⓑ Ⓒ Ⓓ
146 Ⓐ Ⓑ Ⓒ Ⓓ
147 Ⓐ Ⓑ Ⓒ Ⓓ
148 Ⓐ Ⓑ Ⓒ Ⓓ
149 Ⓐ Ⓑ Ⓒ Ⓓ
150 Ⓐ Ⓑ Ⓒ Ⓓ

DARKEN CIRCLE TO HAVE QUESTIONS 151–160 SCORED
151 Ⓐ Ⓑ Ⓒ Ⓓ
152 Ⓐ Ⓑ Ⓒ Ⓓ
153 Ⓐ Ⓑ Ⓒ Ⓓ
154 Ⓐ Ⓑ Ⓒ Ⓓ
155 Ⓐ Ⓑ Ⓒ Ⓓ
156 Ⓐ Ⓑ Ⓒ Ⓓ
157 Ⓐ Ⓑ Ⓒ Ⓓ
158 Ⓐ Ⓑ Ⓒ Ⓓ
159 Ⓐ Ⓑ Ⓒ Ⓓ
160 Ⓐ Ⓑ Ⓒ Ⓓ

DARKEN CIRCLE TO HAVE QUESTIONS 161–170 SCORED
161 Ⓐ Ⓑ Ⓒ Ⓓ
162 Ⓐ Ⓑ Ⓒ Ⓓ
163 Ⓐ Ⓑ Ⓒ Ⓓ
164 Ⓐ Ⓑ Ⓒ Ⓓ
165 Ⓐ Ⓑ Ⓒ Ⓓ
166 Ⓐ Ⓑ Ⓒ Ⓓ
167 Ⓐ Ⓑ Ⓒ Ⓓ
168 Ⓐ Ⓑ Ⓒ Ⓓ
169 Ⓐ Ⓑ Ⓒ Ⓓ
170 Ⓐ Ⓑ Ⓒ Ⓓ

DARKEN CIRCLE TO HAVE QUESTIONS 171–180 SCORED
171 Ⓐ Ⓑ Ⓒ Ⓓ
172 Ⓐ Ⓑ Ⓒ Ⓓ
173 Ⓐ Ⓑ Ⓒ Ⓓ
174 Ⓐ Ⓑ Ⓒ Ⓓ
175 Ⓐ Ⓑ Ⓒ Ⓓ
176 Ⓐ Ⓑ Ⓒ Ⓓ
177 Ⓐ Ⓑ Ⓒ Ⓓ
178 Ⓐ Ⓑ Ⓒ Ⓓ
179 Ⓐ Ⓑ Ⓒ Ⓓ
180 Ⓐ Ⓑ Ⓒ Ⓓ

DARKEN CIRCLE TO HAVE QUESTIONS 181–190 SCORED
181 Ⓐ Ⓑ Ⓒ Ⓓ
182 Ⓐ Ⓑ Ⓒ Ⓓ
183 Ⓐ Ⓑ Ⓒ Ⓓ
184 Ⓐ Ⓑ Ⓒ Ⓓ
185 Ⓐ Ⓑ Ⓒ Ⓓ
186 Ⓐ Ⓑ Ⓒ Ⓓ
187 Ⓐ Ⓑ Ⓒ Ⓓ
188 Ⓐ Ⓑ Ⓒ Ⓓ
189 Ⓐ Ⓑ Ⓒ Ⓓ
190 Ⓐ Ⓑ Ⓒ Ⓓ

DARKEN CIRCLE TO HAVE QUESTIONS 191–200 SCORED
191 Ⓐ Ⓑ Ⓒ Ⓓ
192 Ⓐ Ⓑ Ⓒ Ⓓ
193 Ⓐ Ⓑ Ⓒ Ⓓ
194 Ⓐ Ⓑ Ⓒ Ⓓ
195 Ⓐ Ⓑ Ⓒ Ⓓ
196 Ⓐ Ⓑ Ⓒ Ⓓ
197 Ⓐ Ⓑ Ⓒ Ⓓ
198 Ⓐ Ⓑ Ⓒ Ⓓ
199 Ⓐ Ⓑ Ⓒ Ⓓ
200 Ⓐ Ⓑ Ⓒ Ⓓ

Mark Reflex® by NCS EM-221000-5:654321   ED05   Printed in U.S.A.

NATIONAL COUNCIL OF EXAMINERS FOR ENGINEERING & SURVEYING

1 Please PRINT your NAME below:

Last name   First name   Middle Initial

EXAM DATE _____
Month   Day   Year

LOCATION _____
City   State

NCEES

2 EXAMINEE
IDENTIFICATION NUMBER

0 0 0 0 0 0 0 0 0
1 1 1 1 1 1 1 1 1
2 2 2 2 2 2 2 2 2
3 3 3 3 3 3 3 3 3
4 4 4 4 4 4 4 4 4
5 5 5 5 5 5 5 5 5
6 6 6 6 6 6 6 6 6
7 7 7 7 7 7 7 7 7
8 8 8 8 8 8 8 8 8
9 9 9 9 9 9 9 9 9

3 LAST NAME
First 4 letters | 1st. Init.

4 DATE OF BIRTH

MONTH | DAY | 19 YEAR

Jan Feb Mar Apr May June July Aug Sept Oct Nov Dec

0 1 2 3 4 5 6 7 8 9

(letter columns A–Z, first 4 letters of last name and 1st initial)

A A A A A
B B B B B
C C C C C
D D D D D
E E E E E
F F F F F
G G G G G
H H H H H
I I I I I
J J J J J
K K K K K
L L L L L
M M M M M
N N N N N
O O O O O
P P P P P
Q Q Q Q Q
R R R R R
S S S S S
T T T T T
U U U U U
V V V V V
W W W W W
X X X X X
Y Y Y Y Y
Z Z Z Z Z

6 If you are a graduate of an engineering or science related curriculum

19 ____ ____   20 ____ ____

0 1 2 3 4 5 6 7 8 9   0 1 2 3 4 5 6 7 8 9   (and 0,1)

Indicate the last 2 digits of the year you graduated

5 Board from which licensure is sought

Ala.   La.   Ohio
Alaska   Maine   Okla.
Ariz.   Mass.   Oreg.
Ark.   Md.   Pa.
Calif.   Mich.   P.R.
Colo.   Minn.   R.I.
Conn.   Miss.   S.C.
D.C.   Mo.   S. Dak.
Del.   Mont.   Tenn.
Fla.   MP   Tex.
Ga.   N.C.   Utah
Guam   N. Dak.   Va.
Hawaii   Nebr.   V.I.
Idaho   Nev.   Vt.
Illinois   N.H.   Wash.
Ind.   N.J.   W. Va.
Iowa   N. Mex.   Wis.
Kans.   N.Y.   Wyo.
Ky.

DATA FROM THIS PAGE IS GATHERED FOR STATISTICAL PURPOSES AND HAS NO EFFECT ON SCORING.

7 Blacken the circle that best describes your Baccalaureate Degree
I AM A GRADUATE OF:
- 4-year Engineering (ABET accredited)
- 4-year Engineering Technology (ABET accredited)
- 4-year Engineering (not accredited)
- 4-year Engineering Technology (not accredited)
- None of the above
- Do not know

8 I am taking the exam for the:
- First time
- Second time
- Third time
- Fourth time (or more)

9 PRINT INSIDE BOXED-IN AREA

EXAM BOOKLET SERIAL NUMBER
(from top right corner of book)

10 PLEASE READ, THEN SIGN YOUR NAME BELOW:

I affirm by my signature below that I am the person taking this exam, the answers contained hereon are solely of my knowledge and hand, and I have not taken this exam in the previous 30 days.

I further affirm that I will not copy any information onto material to be taken from the exam room. Nor will I reveal in whole or in part any exam questions, answers, problems or solutions to anyone during or after the exam, whether orally, in writing, or any internet "chat rooms," or otherwise. I understand that failure to comply with this statement could result in an invalidation of my exam results and/or bar me from retaking the exam for a time at the discretion of the board.

Signature

USE NO. 2 PENCIL ONLY

**PRINCIPLES & PRACTICE
OF ENGINEERING
EXAMINATION ANSWER SHEET**

SIDE 1   P.M.

Mark Reflex® by NCS EM-159289-8:654321   ED06   Printed in U.S.A.

SAMPLE

FOR MULTIPLE-CHOICE QUESTIONS ONLY
TO BE COMPLETED BY EXAMINEE

DARKEN THE CIRCLE BESIDE THE DISCIPLINE TO BE WORKED

THE FOLLOWING DISCIPLINES OFFER A CHOICE OF QUESTIONS. WORK ONLY 40 QUESTIONS WHICH MUST BE ANSWERED IN 4 SETS OF 10.

○ Electrical   ○ Metallurgical   ○ Mining/Mineral

THE FOLLOWING DISCIPLINES ARE NO-CHOICE. QUESTIONS 501-540 WILL BE SCORED (FOR ENVIRONMENTAL, QUESTIONS 501-550 WILL BE SCORED).

○ Agricultural
○ Chemical
○ Civil-Environmental
○ Civil-Geotechnical
○ Civil-Structural
○ Civil-Transportation
○ Civil-Water Resources
○ Control Systems
○ Environmental
Fire Protection
○ Industrial
○ Manufacturing
○ Mechanical-HVAC & Refrigeration
○ Mechanical-Machine Design
○ Mechanical-Thermal & Fluids Systems
○ Naval Architecture/Marine
○ Nuclear
○ Petroleum
○ Structural I

IDENTIFICATION NUMBER

[ ][ ][ ][ ][ ][ ][ ][ ][ ][ ]

FOR NCEES USE ONLY

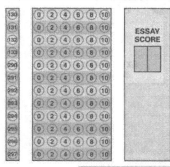

 ○ AM ONLY

○ PM ONLY

 NUMBER OF ESSAY ITEMS WORKED

ESSAY SCORE

DARKEN THE CIRCLES FOR THE ANSWERS BELOW TO THE QUESTIONS YOU WISH TO HAVE SCORED

DARKEN CIRCLE○ TO HAVE QUESTIONS 501-510 SCORED

501 Ⓐ Ⓑ Ⓒ Ⓓ
502 Ⓐ Ⓑ Ⓒ Ⓓ
503 Ⓐ Ⓑ Ⓒ Ⓓ
504 Ⓐ Ⓑ Ⓒ Ⓓ
505 Ⓐ Ⓑ Ⓒ Ⓓ
506 Ⓐ Ⓑ Ⓒ Ⓓ
507 Ⓐ Ⓑ Ⓒ Ⓓ
508 Ⓐ Ⓑ Ⓒ Ⓓ
509 Ⓐ Ⓑ Ⓒ Ⓓ
510 Ⓐ Ⓑ Ⓒ Ⓓ

DARKEN CIRCLE○ TO HAVE QUESTIONS 511-520 SCORED

511 Ⓐ Ⓑ Ⓒ Ⓓ
512 Ⓐ Ⓑ Ⓒ Ⓓ
513 Ⓐ Ⓑ Ⓒ Ⓓ
514 Ⓐ Ⓑ Ⓒ Ⓓ
515 Ⓐ Ⓑ Ⓒ Ⓓ
516 Ⓐ Ⓑ Ⓒ Ⓓ
517 Ⓐ Ⓑ Ⓒ Ⓓ
518 Ⓐ Ⓑ Ⓒ Ⓓ
519 Ⓐ Ⓑ Ⓒ Ⓓ
520 Ⓐ Ⓑ Ⓒ Ⓓ

DARKEN CIRCLE○ TO HAVE QUESTIONS 521-530 SCORED

521 Ⓐ Ⓑ Ⓒ Ⓓ
522 Ⓐ Ⓑ Ⓒ Ⓓ
523 Ⓐ Ⓑ Ⓒ Ⓓ
524 Ⓐ Ⓑ Ⓒ Ⓓ
525 Ⓐ Ⓑ Ⓒ Ⓓ
526 Ⓐ Ⓑ Ⓒ Ⓓ
527 Ⓐ Ⓑ Ⓒ Ⓓ
528 Ⓐ Ⓑ Ⓒ Ⓓ
529 Ⓐ Ⓑ Ⓒ Ⓓ
530 Ⓐ Ⓑ Ⓒ Ⓓ

DARKEN CIRCLE○ TO HAVE QUESTIONS 531-540 SCORED

531 Ⓐ Ⓑ Ⓒ Ⓓ
532 Ⓐ Ⓑ Ⓒ Ⓓ
533 Ⓐ Ⓑ Ⓒ Ⓓ
534 Ⓐ Ⓑ Ⓒ Ⓓ
535 Ⓐ Ⓑ Ⓒ Ⓓ
536 Ⓐ Ⓑ Ⓒ Ⓓ
537 Ⓐ Ⓑ Ⓒ Ⓓ
538 Ⓐ Ⓑ Ⓒ Ⓓ
539 Ⓐ Ⓑ Ⓒ Ⓓ
540 Ⓐ Ⓑ Ⓒ Ⓓ

DARKEN CIRCLE○ TO HAVE QUESTIONS 541-550 SCORED

541 Ⓐ Ⓑ Ⓒ Ⓓ
542 Ⓐ Ⓑ Ⓒ Ⓓ
543 Ⓐ Ⓑ Ⓒ Ⓓ
544 Ⓐ Ⓑ Ⓒ Ⓓ
545 Ⓐ Ⓑ Ⓒ Ⓓ
546 Ⓐ Ⓑ Ⓒ Ⓓ
547 Ⓐ Ⓑ Ⓒ Ⓓ
548 Ⓐ Ⓑ Ⓒ Ⓓ
549 Ⓐ Ⓑ Ⓒ Ⓓ
550 Ⓐ Ⓑ Ⓒ Ⓓ

DARKEN CIRCLE○ TO HAVE QUESTIONS 551-560 SCORED

551 Ⓐ Ⓑ Ⓒ Ⓓ
552 Ⓐ Ⓑ Ⓒ Ⓓ
553 Ⓐ Ⓑ Ⓒ Ⓓ
554 Ⓐ Ⓑ Ⓒ Ⓓ
555 Ⓐ Ⓑ Ⓒ Ⓓ
556 Ⓐ Ⓑ Ⓒ Ⓓ
557 Ⓐ Ⓑ Ⓒ Ⓓ
558 Ⓐ Ⓑ Ⓒ Ⓓ
559 Ⓐ Ⓑ Ⓒ Ⓓ
560 Ⓐ Ⓑ Ⓒ Ⓓ

DARKEN CIRCLE○ TO HAVE QUESTIONS 561-570 SCORED

561 Ⓐ Ⓑ Ⓒ Ⓓ
562 Ⓐ Ⓑ Ⓒ Ⓓ
563 Ⓐ Ⓑ Ⓒ Ⓓ
564 Ⓐ Ⓑ Ⓒ Ⓓ
565 Ⓐ Ⓑ Ⓒ Ⓓ
566 Ⓐ Ⓑ Ⓒ Ⓓ
567 Ⓐ Ⓑ Ⓒ Ⓓ
568 Ⓐ Ⓑ Ⓒ Ⓓ
569 Ⓐ Ⓑ Ⓒ Ⓓ
570 Ⓐ Ⓑ Ⓒ Ⓓ

DARKEN CIRCLE○ TO HAVE QUESTIONS 571-580 SCORED

571 Ⓐ Ⓑ Ⓒ Ⓓ
572 Ⓐ Ⓑ Ⓒ Ⓓ
573 Ⓐ Ⓑ Ⓒ Ⓓ
574 Ⓐ Ⓑ Ⓒ Ⓓ
575 Ⓐ Ⓑ Ⓒ Ⓓ
576 Ⓐ Ⓑ Ⓒ Ⓓ
577 Ⓐ Ⓑ Ⓒ Ⓓ
578 Ⓐ Ⓑ Ⓒ Ⓓ
579 Ⓐ Ⓑ Ⓒ Ⓓ
580 Ⓐ Ⓑ Ⓒ Ⓓ

DARKEN CIRCLE○ TO HAVE QUESTIONS 581-590 SCORED

581 Ⓐ Ⓑ Ⓒ Ⓓ
582 Ⓐ Ⓑ Ⓒ Ⓓ
583 Ⓐ Ⓑ Ⓒ Ⓓ
584 Ⓐ Ⓑ Ⓒ Ⓓ
585 Ⓐ Ⓑ Ⓒ Ⓓ
586 Ⓐ Ⓑ Ⓒ Ⓓ
587 Ⓐ Ⓑ Ⓒ Ⓓ
588 Ⓐ Ⓑ Ⓒ Ⓓ
589 Ⓐ Ⓑ Ⓒ Ⓓ
590 Ⓐ Ⓑ Ⓒ Ⓓ

DARKEN CIRCLE○ TO HAVE QUESTIONS 591-600 SCORED

591 Ⓐ Ⓑ Ⓒ Ⓓ
592 Ⓐ Ⓑ Ⓒ Ⓓ
593 Ⓐ Ⓑ Ⓒ Ⓓ
594 Ⓐ Ⓑ Ⓒ Ⓓ
595 Ⓐ Ⓑ Ⓒ Ⓓ
596 Ⓐ Ⓑ Ⓒ Ⓓ
597 Ⓐ Ⓑ Ⓒ Ⓓ
598 Ⓐ Ⓑ Ⓒ Ⓓ
599 Ⓐ Ⓑ Ⓒ Ⓓ
600 Ⓐ Ⓑ Ⓒ Ⓓ

DARKEN CIRCLE○ TO HAVE QUESTIONS 601-610 SCORED

601 Ⓐ Ⓑ Ⓒ Ⓓ
602 Ⓐ Ⓑ Ⓒ Ⓓ
603 Ⓐ Ⓑ Ⓒ Ⓓ
604 Ⓐ Ⓑ Ⓒ Ⓓ
605 Ⓐ Ⓑ Ⓒ Ⓓ
606 Ⓐ Ⓑ Ⓒ Ⓓ
607 Ⓐ Ⓑ Ⓒ Ⓓ
608 Ⓐ Ⓑ Ⓒ Ⓓ
609 Ⓐ Ⓑ Ⓒ Ⓓ
610 Ⓐ Ⓑ Ⓒ Ⓓ

DARKEN CIRCLE○ TO HAVE QUESTIONS 611-620 SCORED

611 Ⓐ Ⓑ Ⓒ Ⓓ
612 Ⓐ Ⓑ Ⓒ Ⓓ
613 Ⓐ Ⓑ Ⓒ Ⓓ
614 Ⓐ Ⓑ Ⓒ Ⓓ
615 Ⓐ Ⓑ Ⓒ Ⓓ
616 Ⓐ Ⓑ Ⓒ Ⓓ
617 Ⓐ Ⓑ Ⓒ Ⓓ
618 Ⓐ Ⓑ Ⓒ Ⓓ
619 Ⓐ Ⓑ Ⓒ Ⓓ
620 Ⓐ Ⓑ Ⓒ Ⓓ

Sample questions and solutions are also available for the following examinations:

Civil Engineering
Chemical Engineering
Environmental Engineering
Mechanical Engineering
Structural I Engineering
Structural II Engineering

For more information about these and other Council study materials (including CD-ROMs and Internet Diagnostic tests), visit our homepage at www.ncees.org or contact our Customer Service Department at 800-250-3196.